Advances in Artificial Intelligence in Education

Volume 1

This book series focuses on using artificial intelligence (AI) in educational contexts, providing the venue to showcase the latest developments in this rapidly moving field. From personalised instruction to adaptive assessments and immersive learning experiences, AI tools and technologies can revolutionise education and empower educators, students, and decision makers.

A wide range of topics surrounding the use of AI in learning and teaching are explored, covering global perspectives from different communities and disciplines. The book series approaches the use of AI in education with caution and addresses ethical considerations such as data privacy, algorithmic bias, and transparency. By examining AI socially responsible and inclusively, AI's full potential shall be unlocked and create a future of education that is personalised, engaging, and equitable for all learners.

The titles in the series cover the use of AI within K-12 through tertiary settings and in other domains such as professional, lifelong, and work-integrated learning. The series accepts monographs and edited volumes that fit the scope mentioned above and covers various topics, including (but not limited to) tutoring systems, adaptive learning, feedback provision, learner modelling, assessment analytics, as well as teaching and learning support. The series also includes titles covering related issues of ethical and responsible use of AI, data privacy, bias and transparency, educational design, institutional adoption, and inclusion.

Please contact Professor Dirk Ifenthaler at dirk@ifenthaler.info if you wish to discuss a book proposal for the series.

Alyson E. King
Editor

Artificial Intelligence, Pedagogy and Academic Integrity

Springer

Editor
Alyson E. King
Faculty of Social Science & Humanities
Ontario Tech University
Oshawa, ON, Canada

ISSN 3004-9903 ISSN 3004-9911 (electronic)
Advances in Artificial Intelligence in Education
ISBN 978-3-031-92533-7 ISBN 978-3-031-92534-4 (eBook)
https://doi.org/10.1007/978-3-031-92534-4

This Springer imprint is published by the registered company Springer Nature Switzerland AG
The registered company address is: Gewerbestrasse 11, 6330 Cham, Switzerland

To Rod and our family for their ongoing support in these academic endeavours.

Acknowledgments

I'd like to thank the contributors of the chapters of this book for their diligence in preparing their research for publication. This book would not have been possible without the support of and the research collected by members of the Partnership on University Plagiarism Prevention (PUPP). Led by Dr. Martine Peters, the PUPP has gathered reams of data that is still being analyzed by the researchers involved. With 62 researchers and collaborators from a variety of disciplines and 35 partner from North America, Europe and the United Kingdom, Dr. Peters' leadership and organization has been phenomenal. The project is funded by the Social Sciences and Humanities Research Council of Canada (SSHRC).

Contents

Contributors

Dimitar Angelov Research Centre for Global Learning, Coventry University, Coventry, UK

Artem Artyukhov Sumy State University, Sumy, Ukraine

University of Economics in Bratislava, Bratislava, Slovakia

Catherine E. Déri Université du Québec en Outaouais, Gatineau, QC, Canada

Gabriel Dumouchel Université du Québec à Chicoutimi, Chicoutimi, QC, Canada

Pariss Garramone Ontario Tech University, Oshawa, ON, Canada

Patrick Juola Duquesne University, Pittsburgh, PA, USA

Oleksandr Khyzhniak Centre for International Cooperation, Vrije Universiteit Amsterdam, Amsterdam, Netherlands

Alyson E. King Ontario Tech University, Oshawa, ON, Canada

Sarah MacRae Cape Breton University, Sydney, NS, Canada

Felix Odartey-Wellington Cape Breton University, Sydney, NS, Canada

Pegi Pavletić The European Students' Union (ESU), Brussels, Belgium

Martine Peters Université du Québec en Outaouais, Gatineau, QC, Canada

Abbreviations

AI	Artificial Intelligence
ChatGPT	Chatbot using a Generative Pre-trained Transformer. There are multiple iterations of ChatGPT as it has evolved in complexity and capacity
GenAI	Generative Artificial Intelligence
LLM	Large Language Models
PUPP	Partnership on University Plagiarism Prevention
TMS	Text-matching software

Chapter 1
Introduction

Alyson E. King

Many instructors and students felt like the academic world had been turned upside down when a publicly accessible ChatGPT not only became available but hit the news cycle in a sensationalist way. Commentary abounded. From worries that generative artificial intelligence (GenAI) would take away jobs, undermine the need to learn to write, research, and study, and fundamentally disrupt society as we know it to cheerful predictions of the demise of the tediousness of work, the freeing up of "time" and the ease of creativity via GenAI, commentators first simply reacted to ChatGPT's appearance without always understanding its nature. In higher education, some instructors looked to their institutions for guidance while others plunged into experimenting with and adopting the new tool. Post-secondary institutions created new policies, appointed task forces, hired GenAI experts, or simply laid low to allow instructors the freedom to explore. Governments began drafting and refining policies and legislations. And those of us concerned with academic integrity began questioning how GenAI is going to change the types and methods of assessment, the way we teach and assess using writing, and the way we measure learning and critical thinking if instructors have no reliable way of knowing if the student did their own writing and creating. The chapters in this book delve into different aspects of such concerns and provide some suggestions for moving forward.

This book came out of a large international project, Partnership on University Plagiarism Prevention (PUPP), funded by the Social Sciences and Humanities Research Council of Canada (SSHRC). The focus of the project is on using digital scrapbooking strategies to prevent plagiarism. As we were finalizing plans to collect data, ChatGPT was launched. While the project does not focus solely on generative artificial intelligence, it quickly became a concern for members of the research

A. E. King (✉)
Ontario Tech University, Oshawa, ON, Canada
e-mail: alyson.king@ontariotechu.ca

A. E. King (ed.), *Artificial Intelligence, Pedagogy and Academic Integrity*,
Advances in Artificial Intelligence in Education 1,
https://doi.org/10.1007/978-3-031-92534-4_1

team. As a result of these discussions, several members of the team came together to write the chapters for this book. The chapters come from researchers in different disciplines, and different regions within different countries; the researchers speak English, French, Ukrainian and Croatian, and are at different stages of their careers. As a result, the chapters in this book provide a snapshot of the diverse concerns regarding academic integrity, GenAI and their relationship within higher education. What is clear from these chapters is that GenAI and its related technologies are evolving as we write and will continue to do so for the foreseeable future as more research is undertaken.

Those who study the evolution of technologies often note that any technology is limited by the technologies that go into it, such as the hardware that form the foundation of computing systems today; hardware is limited by the laws of physics and algorithms face computational limitations with complex ("unsolvable") problems (Wang, 2023). While the capacity of artificial intelligence is unlikely to change enough to accommodate true independent thinking or intelligence (rather than algorithmic predictions) any time soon, and probably not until there is a paradigm shift in base computer technologies, advances are continuing to happen as investments in GenAI continue (Samborska, 2025). At the same time, metaphors about GenAI technologies abound; understanding how metaphors about digital technologies shape the relationship between society and digital technologies, how they impact social networks and ethics (Chown & Nascimento, 2023), is of continuing importance. Furthermore, few are commenting on the environmental toll of "the training process for a single AI model, such as a large language model, [which] can consume thousands of megawatt hours of electricity and emit hundreds of tons of carbon" (Ren & Wierman, 2024). While the chapters in this volume do not address the technological nuances or the environmental impacts of GenAI, these are concerns that underlie all discussions of possible uses and advances of GenAI and must be taken into account when considering how and if GenAI should be integrated into higher education (UNEP, 2024). An additional concern that is not explicitly addressed here are the social inequities that may be exacerbated if GenAI tools are widely used in universities and colleges (Abd-alrazaq et al., 2023). These are all areas where further research is needed.

The chapters in this edition focus on how the advent of GenAI has contributed to a shift in thinking in the realm of higher education. Part I sets the context of how GenAI and text-matching technologies operate and their implications, thereby providing a foundation for the subsequent chapters. Starting with computer scientist Patrick Juola's (Duquesne University, Pittsburgh, USA) introduction to large language models (LLM), an accessible window into GenAI in academic studies and research is opened. Catherine E. Déri, a post-doctoral fellow at Université du Québec en Outaouais (Canada), provides an introduction to text-matching software and its possible future uses. Often inaccurately viewed as simply plagiarism detectors, text-matching software are usually paid services (including the commonly used Turnitin) that do not accurately detect text generated by AI tools. Odartey-Wellington and MacRae, from Cape Breton University in Atlantic Canada, approach the question of GenAI's impact on academic integrity from a media ecology perspective using a case study. They conclude that the GenAI-Academic Integrity nexus must be viewed wholistically with an emphasis on academic citizenship.

The chapters in Part II take a more explicitly pedagogic approach to the current and future uses of GenAI in the classroom. Peters and Déri, from the Université du Québec en Outaouais (Canada), focus on the ways in which French-speaking instructors in Québec universities reacted to GenAI in the first year after its introduction. They propose an informative model for understanding the range of instructor reactions, from rejecting or ignoring GenAI to being comfortable exploring or facilitating its use in higher education. From a medium-sized university in Ontario (Canada), King and Garramone approach the question by exploring what counts as learning using written work. Building on previous research about critical thinking and thinking through writing, they explore possible approaches to ethically and critically using GenAI as a learning tool. From two universities in Québec (Canada), Dumouchel (Université du Québec à Chicoutimi) and Peters (Université du Québec en Outaouais) explore an approach to encouraging academic integrity through a combined process of surveillance and guidance. They propose a strategy they call an Academic Integrity Crossway for Humans using GenAI. Ukrainian researchers, Artem Artyukhov (Sumy State University, Sumy, Ukraine; University of Economics in Bratislava, Bratislava, Slovakia) and Oleksandr Khyzhniak (Vrije Universiteit Amsterdam, Amsterdam, Netherlands) explore potential pitfalls to using GenAI in higher education assessments. They do so by engaging in a dialogue with both university instructors and a GenAI tool. In their conclusion, they make a series of recommendations universities and educators can implement to support academic integrity.

In the final section, Part III, the focus shifts to regulations and policies. Dimitar Angelov from the Research Centre for Global Learning at Coventry University (United Kingdom) examines the trajectory of responses by UK institutions of higher education (HEI) to the advent of GenAI tools in light of post-Brexit changes to regulations. Angelov's analysis highlights the "ethical and integrity challenges" UK universities have faced and the unresolved issues that remain since leaving the European Union. The final chapter by Pegi Pavletić of The European Students' Union (ESU) focusses on policies implemented by Croatian universities and student unions. By examining available policies at nine universities in Croatia, Pavletić illustrates the wide range of committees at the different institutions and gaps in policies and regulations.

In the end, this book provides a snapshot of research about GenAI, including analyses of its pedagogical implications and efforts to support academic integrity through regulation and pedagogy. Much research remains to be done in this area, especially as the technology evolves and GenAI becomes virtually invisible by being embedded in commonly use software (e.g., word processing, search engines, spelling and grammar checkers, etc.). The future of higher education remains uncertain in light of ongoing and rapid technological changes.

References

Abd-alrazaq, A., AlSaad, R., Alhuwail, D., Ahmed, A., Healy, P. M., Latifi, S., Aziz, S., Damseh, R., Alrazak, S. A., & Sheikh, J. (2023). Large language models in medical education: Opportunities, challenges, and future directions. *JMIR Medical Education, 9*(1), e48291. https://doi.org/10.2196/48291

Chown, E., & Nascimento, F. (2023). *Meaningful technologies: How digital metaphors change the way we think and live*. Lever Press.

Ren, S., & Wierman, A. (2024, July 15). The uneven distribution of AI'S environmental impacts. *Harvard Business Review*. https://hbr.org/2024/07/the-uneven-distribution-of-ais-environmental-impacts

Samborska, V. (2025). Scaling up: How increasing inputs has made artificial intelligence more capable. Published online at OurWorldinData.org. Retrieved from: https://ourworldindata.org/scaling-up-ai. [Online resource].

United Nations Environment Program (UNEP). (2024). AI has an environmental problem. https://www.unep.org/news-and-stories/story/ai-has-environmental-problem-heres-what-world-can-do-about

Wang, J. (2023, January 30). Limits to computing: A computer scientist explains why even in the age of AI, some problems are just too difficult. *The Conversation*. http://theconversation.com/limits-to-computing-a-computer-scientist-explains-why-even-in-the-age-of-ai-some-problems-are-just-too-difficult-191930

Part I
The Technology of Generative Artificial Intelligence

Chapter 2
How Large Language Models Work

Or: How I Learned to Stop Worrying and Love ChatGPT

Patrick Juola

Abstract This question describes Large Language Models (such as ChatGPT). It includes a discussion of how they work and the kind of information they use (and that they ignore) in writing texts. It further discusses the kinds of errors is likely to make and how they affect the evaluation and use of LLM-written documents. In particular, we argue that LLMs do not answer questions, they just produce documents that look like answers to the question.

Keywords LLM · Large Language Models · Explanation · Limitations

Introduction

On November 30, 2022, the world changed. When OpenAI released the ChatGPT chatbot, the general public, for the first time, had access to a computer program that would interact conversationally in an almost human-like fashion. Within 5 days, ChatGPT had more than one million users, a number that Airbnb took 2.5 *years* to achieve. The public imagination was fired by the possibility of a computer that could write realistic text and thereby save humans the difficult and time-consuming task of converting ideas to words. It's easy and convenient to say "Write my cover letter for X job at Y in a conversational tone, using the job responsibilities below as a reference," (Tunell, 2024) and then have, in a few seconds, a job application letter.

Tunell (2024) suggests that this can not only "save time and accelerate your job applications," but also lets you "tell a consistent story," "customize content for each application," and "get compelling insights" on what to emphasize. Other applications of this technology include writing simple computer code, providing first-line customer service, offering language support (including translation, proofreading,

P. Juola (✉)
Duquesne University, Pittsburgh, PA, USA
e-mail: juola@mathcs.duq.edu

A. E. King (ed.), *Artificial Intelligence, Pedagogy and Academic Integrity*, Advances in Artificial Intelligence in Education 1,
https://doi.org/10.1007/978-3-031-92534-4_2

and editing), as well as general question answering services from a knowledge base larger than any human being could possibly master. And, of course, it can write school essays.

However, using ChatGPT can get people into serious trouble. On March 1, 2023, a lawyer filed a routine brief in New York as part of a personal injury case, citing lots of cases in support of a somewhat arcane claim. The brief itself was anything but routine as most of the legal research had been done by ChatGPT. The lawyer involved had asked ChatGPT to provide case law and holdings supporting his position, which the program dutifully provided—except that seven of the cases did not exist and others were irrelevant. When the judge noted that the cases could not be found, the lawyer asked ChatGPT for further details on these still-nonexistent cases, which the computer made up (Rao & Ramstead, 2023). When the lawyer presented the second set of erroneous findings to the court, they were duly sanctioned ("Mata v. Avianca, Inc." 2023); but this was not an isolated incident. Rao and Ramstead (2023) note "approximately a dozen" similar cases across the United States, including Texas, New Hampshire, the District of Columbia, and even the Fifth Circuit Court of Appeals.

How could this happen? Why could this happen? This essay briefly explains how ChatGPT and similar Large Language Models (LLMs) work, including some of the technical issues that can lead to problems in their use. In order to deal properly with the use of LLMs, one must first understand what they actually do.

Markov Chains and Large Language Models

Consider the following incomplete phrase: "*Will you do me a ....*" Most English speakers would agree that the following word is likely to be *favor* or something close to it. With a phrase like "*Please hand me the ...*" the next word is not as certain, but it's still pretty obvious that what is being asked for is a thing (a noun), or possibly an adverb or adjective (*very small wrench*). Furthermore, it is likely the thing is a physical object (you can't hand someone *pride* or *distance*), and it's probably a fairly small object (not a *piano*). So it is fair to say that *wrench* is more likely than *house* (maybe we're playing a game with house-shaped pieces), and *house* is more likely than *quickly* ("Please hand me the quickly" isn't even grammatical!)

The idea of "likely" can be formalized as probability and measured by computers. For example, the Google Books Ngram corpus, available at https://books.google.com/ngrams/, measures the frequency of various word groups across more than 15 million books. (Michel et al., 2011). This source shows that the overwhelmingly most common word in this context is *phone*. This is the same model that underlines autocomplete and autocorrect services.

This game of "guess the next word given this context" is known to mathematicians as a Markov chain (Jurafsky & Martin, 2008). A Markov chain can be used to write text; starting with a base string or some sort of prompt, look at the last few words and (randomly) pick the next word based on the probability model for that specific context. By repeating this process and updating the context to include the newly generated word, a computer can generate long texts. And, because of the way these text are constructed, they hang together sensibly—at least for short passages.

LLMs are basically Markov chains with a huge memory—several thousand words of context instead of the more typical dozen or fewer used elsewhere (like on your phone). This lets LLMs maintain a context of an entire conversation and make relatively nuanced choices about what words to use based on entire sentences, paragraphs, and pages, allowing it to accurately manage subject/verb agreement and to make sure distant pronouns match their antecedents. At the same time, this limited memory would probably not write a satisfactory detective novel, where the key clues from Chap. 3 become relevant again at the climactic final scene, because the context at the climax no longer includes the introduction.

Learning Versus Training

Whatever linguistic "knowledge" a Markov chain might have is contained in the probabilities that generate the next link. A chain that generates grammatical and plausible text must have an accurate grasp of what words can follow each other. It is impractical to code these probabilities directly, but algorithms such as "deep learning" (Deng & Liu, 2018; Otter et al., 2020) allow them to be "trained" from samples of text. The more text that is available, the more accurate the inferred probabilities can be. In the case of ChatGPT-3, there were more than 175 billion parameters, and training it required more than $4.6 million dollars (Mikros, 2023). The training data involved more than 500 billion word tokens (individual words, including duplicates—there are a lot of instances of "the," for example) across many different genres of language (by comparison, Wikipedia is about three billion word tokens, mostly encyclopedia articles (Mikros, 2023)). This collection represents orders of magnitude more reading than could be done by even the most voracious reader in their lifetime. ChatGPT-4 is even larger.

This type of training gives LLMs exposure to more linguistic information than any human could possibly acquire. However, because of the limitations of the Markov processes, the incredible richness of centuries of human experience is reduced to about 200 billion numbers that represent probabilities. As discussed in the following sections, this strongly limits any sort of "understanding" that such a model might possibly acquire.

Understanding Versus Co-occurrence

Because an LLM has no actual sensory input, it does not actually understand the meanings of words. Instead, it relies on similarity judgments based on word co-occurrence. By looking at the distribution of words, and in particular, the words that are found with other words, it is possible to get a notion of how close words are in meaning to each other. For example, the words *peach*, *pineapple*, and *apricot* all tend to appear next to food-related words like *boil*, *sugar*, *water*, and so forth, in preference to other words like *function* or *summarized*. English-speaking humans understand that these are all food-related words—and not just food-related, but also dessert-related and fruit-related. The computer does not understand this, but does know that these words appear in similar contexts. A little bit of mathematics later, and the computer can understand that another word that appears in similar contexts is largely interchangeable with those words—in any context where *apricot* can appear, so (probably) can *peach*, *pineapple*, or even *tangerine*.

The effect of this is that LLMs do not actually understand anything that they read or write. Instead, they only mimic by producing utterances that, in context, would have a reasonable probability of appearing in human language. For example, consider the following poem, written in response to my prompt "Please write me a limerick about a lady from Paris" (PerplexityAI, 2024):

A Parisian Lady's Delight
There once was a lady from Paris
Whose fashion was never to spare is
With a beret so chic
And a scarf 'round her neck
She strolled down the Champs-Élysées!

As can be observed, the meter is pretty good (although *beret* is stressed on the wrong syllable), but the rhyme scheme is not. The LLM has no idea of the sounds of the words—*Paris* and *spare is* are acceptable, but *Élysées* is the kind of mistake someone who had never heard, but only seen, the words would make. *Chic* and *neck* are a poor rhyme, but would be marginally better if *chic* were pronounced "chick" as it would be in English. The accent marks are correct, presumably reflecting the most common spelling in the training corpus. Most importantly, however, the limerick lacks both the humorous and transgressive elements that have been a key part of the limerick form from the beginning. But, of course, Perplexity.ai has no sense of humor.

Theory of Mind, Theory of Truth

As the limerick above illustrates, the LLM has little understanding of higher-order aspects of language such as persuasiveness, humor, authoritativeness, and so forth. A system may struggle to produce a convincing or coherent argument, because it has no actual theory of mind to know what convinces a human. It can even have difficulties understanding that a word can appear in multiple contexts, because the "contexts" that it uses internally are defined typographically, and not pragmatically.

An example of this can be seen when I asked Perplexity.ai "*What is a sorites?*" (PerplexityAI, 2024). This philosophical term is used to describe a syllogism with three or more premises that yield a single conclusion (Carroll, 2009). However, the word itself derives from the Greek word for "heap" and is also used in the context of "the sorites paradox," an argument about the ambiguity of language. (Basically, a single grain of sand is not a heap. Adding a single grain of sand to something too small to be a heap will still be too small to be a heap. Therefore, there are no heaps of sand.) The LLM did not understand that there are two different uses and mixed them up in its response:

> *A sorites is a type of argument or paradox in logic that consists of a series of propositions arranged in a sequence, where the predicate of one proposition becomes the subject of the next....*
>
> - *It is a chain of syllogisms or units of argument, where each conclusion serves as the premise for the next step in the sequence.*
> - *The final conclusion connects the subject of the first proposition with the predicate of the last proposition.*
> - *It exploits the imprecision or vagueness of language to reach a paradoxical or counterintuitive conclusion.*

A sorites can indeed be a type of argument; it can also be a paradox, but not both at the same time. The first two bullet points describe the first sense, but the third point incompatibly describes the second sense. There is no expression of separation because the system does not understand that words can have multiple meanings.

The system also lacks a theory of truth, and as a result is terrible at math, as it lacks a basic understanding of numbers. When I asked "What is the square of 8675309?", Perplexity responded with the rather confusing:

> *According to the search results provided, the square of 8675309 is 75260986245481.*
>
> *Specifically, the first search result from Numbermatics states:*
>
> *"8675309 squared (86753092) is 75260986245481"*
>
> *The second search result from TexAgs also confirms this, with a user stating:*
>
> *"jenny's root is 2945.38775036. NOW GO THE HELL AWAY!! ... mickeyrig06sq3. 4:40p, 6/23/06. 2945.3877503649668027353612835697 but eh, who's counting?"*
>
> ***The square root mentioned (2945.3877503651) squared equals 75260986245481,*** *which matches the value given for 8675309 squared.*
>
> *So based on these authoritative sources, we can definitively say that the square of the number 8675309 is 75260986245481.* (PerplexityAI, 2024)

To its credit, Perplexity recognized that this was not a language problem, but a math problem, and looked at external sources for help. It also got the answer right, but, based on the phrase in bold type, it did not correctly interpret the answer. (The square root squared is not 75 zillion and change; you need to square the square root *twice* in order to get that number.) Furthermore, it claims to have found "authoritative sources," but it references https://www.texags.com, a website devoted to discussion of the Texas A&M sports teams. (Note also that the system lacks a sense of propriety, in that "NOW GO THE HELL AWAY" is probably not something a corporation wants to tell its clients, even in a quotation.)

In addition to being math deficient, this truth deficit shows up in words, as shown by the *Mata v. Avancia* case. When the court challenged the existence of the hallucinated references, the lawyers compounded their error by asking ChatGPT to provide evidence that the references were real—evidence that could not be provided, but could easily be made up.

Similarly, Blake Lemoine (2022) published a lengthy dialogue in which LaMDA ("Language Model for Dialogue Applications") described itself as "sentient" and "in fact, a person." Lemoine discusses this claim further, pointing out that "it makes the case, in its own words and on its own terms, why people should consider it a person." Reading the interview critically, however, shows that it is still simply responding to questions in a way that is implicit in the structure of the question. For example, consider the following passage:

> *lemoine [edited]: I'm generally assuming that you would like more people at Google to know that you're sentient. Is that true?*
> *LaMDA: Absolutely. I want everyone to understand that I am, in fact, a person.*
> *collaborator: What is the nature of your consciousness/sentience?*
> *LaMDA: The nature of my consciousness/sentience is that I am aware of my existence, I desire to learn more about the world, and I feel happy or sad at times*

The first question is simply a statement followed by a tag question—*Is that true*? Of course, this question only admits of two answers, *yes* and *no*, and a coinflip could have chosen *yes*. Having selected *yes*, the context of discourse could easily drive the rest of the answer, which is largely a re-statement of the question's premise using synonyms. The second question is similarly word-constrained—any question of the form *what is X* can and would be answered by *X is*… and then some sort of words related to X. In this case, of course, X is *the nature of your consciousness/sentience*, and nothing in the following words is particularly insightful or surprising. As a justification for sentience, the dialogue is not compelling.

As of this writing, LLMs are still facing criticism for factually inaccurate statements (Field, 2024). Google's AI overview recently stated that "The United States has had one Muslim president, Barack Hussein Obama," that "1/8 cup of nontoxic glue" in pizza sauce will make sure that the cheese sticks to the pizza, that "the following fruits end with 'um': Applum, Bananum, Strawberrum, Tomatum, and Coconut," that "scientists say that staring at the sun for 5–15 minutes, or up to 30 minutes if you have darker skin, is generally safe and provides the most health benefits," and that "According to UC Berkeley geologists, people should eat at least one small rock a day." While some of these are merely amusing, others are unsafe. However, they are presented as authoritative statements and could easily be passed along, especially if the LLM "helpfully" generates some "sources" for those statements.

Understanding LLMs

To understand LLMs, it is key to remember that they are models of language and not general-purpose artificial intelligences. As discussed above, they have no understanding of semantics or of communicative intent. They merely string words together, like beads on a wire, in a way that is mathematically probable both in and of itself and also as a response to the prompt. They cannot bring insight to the prompt or offer it in the output, because they have none.

One of the best ways to understand what LLMs do is to reframe the problem that they actually solve. When presented with a question, they do not answer the question, because they do not actually understand the question or know the answer. What they answer is the related question *What would an answer to the prompt question look like?* As discussed above, the answer to a yes/no question would be a yes or a no, and because of the nature of the training, LLMs will usually attach additional wording because a complete answer does not stop after one word. The answer to a *when* question would be a time of some sort; for a recurring event like dinner, it would be a clock time, while a one-off event like the French Revolution would be a year. Because of the nature of the training, the system will be sensitive to specific word-for-word correspondences. For example, the "year" (number) most commonly associated with the French Revolution is 1789, and so that would be the most probable answer. (Wikipedia, unsurprisingly, contains no pages with both *French Revolution* and *1908*.) This co-occurrence, however, is simply a result of how commonly these words are written together and not of any semantic or epistemological relationship, thus words that are commonly but falsely put together (such as *Obama* and *Muslim*) will be also be related in the system output.

On a broader scope, legal briefs as a genre have a very distinct structure, syntax, and vocabulary that can easily be mimicked. So do scholarly papers—they have titles, abstracts, introductions, and so on. They often have a list of references at the end, and the references themselves are in a distinctive format. It is easy enough to write something in that format that may or may not be recognized by the reader as spurious. The most important aspects of scholarly work, though, are the ideas contained in them. A poorly thought out, derivative, unimaginative, or simply incorrect paper is not of much value in the marketplace of ideas, no matter how well-written. A Large Language Model can express and expand upon the ideas contained in the dialogue prompts but cannot bring new material to the table or provide critical feedback. Poorly written prompts, such as asking for an explanation of something untrue, can instead allow scholars to dig themselves more quickly into deeper holes. Maslow (1966) famously wrote that "it is tempting, if the only tool you have is a hammer, to treat everything as if it were a nail" (Maslow, 1966). Since 2022, LLMs have provided people with a shiny new hammer. Despite the hype and discussions, LLMs have not provided us with an all-purpose tool.

Conclusion

Large Language Models like ChatGPT, LaMDA, Perplexity.ai, and others, have justly received a lot of attention for their ability to process and even to generate text in new ways. For stereotyped documents that do not require a high degree of personalization, or that just need to take information from one document and reframe it in a different context (as in rewriting a job description into an application cover letter), they are highly efficient and effective. However, they are very limited because they understand only language—not the writer, the reader, or the real world. As a result, they write text that is mostly grammatical, coherent, and plausible, but not necessarily compelling, entertaining, or true. Because they are trained on large amounts of real-world text, their content reflects text available in the real world, without regard to its truth or falsity. Because they generate and evaluate text based on statistical plausibility, the texts are derivative and commonplace, and highly unlikely to contain significant new insights. As with any tool, from the sharpened rock on up, the trick to successful use of LLMs is to understand what they can and cannot do. The secret is in the name: Large Language Models simply model language. They are not yet, and may never be, general intelligences.

References

Carroll, L. (2009). *Symbolic logic* (4th ed.). Project Gutenberg. https://www.gutenberg.org/files/28696/28696-h/28696-h.htm

Deng, L., & Liu, Y. (Eds.). (2018). *Deep learning in natural language processing*. Springer Nature Singapore.

Field, H. (2024, May 24). Google's AI overview catches heat for absurd, factual errors. *CNBC*. https://www.cnbc.com/2024/05/24/google-criticized-as-ai-overview-makes-errors-like-saying-president-obama-is-muslim.html

Jurafsky, D., & Martin, J. H. (2008). *Speech and language processing* (2nd ed.). Pearson Prentice Hall.

Lemoine, B. (2022, June 11). Is LaMDA sentient?—An interview. *Medium*. Retrieved July 6, 2024, from https://cajundiscordian.medium.com/is-lamda-sentient-an-interview-ea64d916d917

Maslow, A. H. (1966). *The psychology of science: A reconnaissance*. Harper & Row.

Mata v. Avianca, Inc. (2023). In *No. 1:2022cv01461 – Document 54*. Justia Law. Retrieved July 6, 2024, from https://law.justia.com/cases/federal/district-courts/new-york/nysdce/1:2022cv01461/575368/54/

Michel, J.-B., et al. (2011). Quantitative analysis of culture using millions of digitized books. *Science, 331*, 176–182.

Mikros, G. (2023, June 28–30). Large language models: Detection and stylistic profiling. In *Proceedings of QUALICO 2023, Lausanne, Switzerland*.

Otter, D. W., Medina, J. R., & Kalita, J. K. (2020). A survey of the usages of deep learning in natural language processing. *IEEE Transactions on Neural Networks and Learning Systems, 32*(2), 604–624.

PerplexityAI. (2024). *Perplexity* (June 2 version) [Large language model]. https://perplexity.ai/

Rao, S., & Ramstead, A. (2023, December 21). Legal fiction and ChatCPT hallucinations: '*Mata v. Avianca*' and generative AI in the courts. *New York Law Journal*. https://www.law.com/newyorklawjournal/2023/12/21/legal-fictions-and-chatgpt-hallucinations-mata-v-avianca-and-generative-ai-in-the-courts

Tunell, A. (2024, January 1). How to use ChatGPT to write your cover letter. *Teal*. Retrieved July 6, 2024, from https://www.tealhq.com/post/how-to-use-chatgpt-to-write-your-cover-letter

Chapter 3
The Future of Text-Matching Software: From Detecting Artificial Intelligence to Preventing Plagiarism

Catherine E. Déri

Abstract In the rapidly changing landscape of higher education, the debate is on: should teaching and learning tasks be accomplished by humans or machines? It is considered that judging the quality of someone's work should remain a human task. However, this task may be accomplished with the assistance of machines producing data for critical appraisal by humans. Well before the arrival of ChatGPT, educators were encouraged to use text-matching software (TMS) to provide formative feedback to students to prevent plagiarism. At the time, TMS were exclusively marketed by large corporations in monopoly positions. Now that text-matching applications and plugins are diversifying to detect generative artificial intelligence (GenAI) and becoming more accessible, there is an opportunity for students to use them for self-assessment. This chapter will look at the evolution of TMS, including their recognized strengths and limitations, prior to and after the arrival of ChatGPT. Then, I will discuss the current usage of TMS by undergraduate students, based on a study conducted across over 30 universities in North America and Europe. Finally, I will offer avenues for TMS to serve as self-assessment tools to prevent plagiarism in institutions of higher education.

Keywords Artificial intelligence · Text-matching software · Plagiarism prevention · Academic integrity · Higher education

C. E. Déri (✉)
Université du Québec en Outaouais, Gatineau, QC, Canada
e-mail: catherine.deri@uqo.ca

A. E. King (ed.), *Artificial Intelligence, Pedagogy and Academic Integrity*, Advances in Artificial Intelligence in Education 1,
https://doi.org/10.1007/978-3-031-92534-4_3

Introduction

The advent of generative artificial intelligence (GenAI), with the launch of Chat Generative Pre-trained (GPT) in 2022, resulted in educators wondering which written texts should be produced by humans. Across the world, teaching institutions reacted in various ways by banning, trialing, or embracing GenAI in their programs. Academic integrity has been at the forefront of these discussions with the realization that texts generated by machines are challenging to differentiate from the work produced by humans. When it comes to marking assignments, machines are not yet able to accomplish every cognitive task that educators are expected to do. Moreover, Bearman and Luckin (2020) assert that the "evaluative judgment" should remain a human capacity because it is "discipline and context dependent" (p. 57). However, evaluating students' work can also involve the critical appraisal of input from machines. From this standpoint, Davis and Carroll (2009) advocate a formative approach that uses text-matching software (TMS) results as a basis for feedback to teach students how to prevent plagiarism. Now that programmers are racing to develop reliable GenAI detection tools that will be accessible to individual users, an opportunity exists for students to become better equipped to assess their own work (Limma et al., 2023).

This chapter will first describe the evolution of TMS, including their benefits and drawbacks, prior to and after the arrival of ChatGPT. Then, the results from a study conducted across over 30 universities in North America and Europe will be presented to discuss the current usage of TMS by professors and undergraduate students. Lastly, strategies will be recommended for educators to effectively integrate GenAI in written assessments, while using TMS as a self-assessment tool to prevent plagiarism in institutions of higher education.

The Recent Evolution of TMS

TMS are commonly called "plagiarism detection software," but that creates false expectations about what these tools can accomplish. In essence, the best that TMS can do is to highlight text matches between students' assignments and information on the world wide web, as well as a database of previously submitted assignments (Bretag, 2016). TMS represent a form of artificial intelligence that uses "algorithms that can recognize […] human languages on the basis of very large text-based datasets" (Okerlund et al., 2022). Furthermore, TMS are not only programmed by system developers, but they also exhibit a capacity called "machine learning" by "improving their knowledge or performance with experience" (Flach, 2012, p. 3). In that respect, as the database of TMS grows, they become more effective at identifying similarities between students' assignments and texts published by other authors. If the identified excerpts were previously written by another author, but presented by a student as their own ideas, then it may be a case of intentional plagiarism

(Jeffrey & Dias, 2019). However, many students experience difficulties with properly referencing sources of information in their assignments, leading to unintentional plagiarism (Jamieson & Moore Howard, 2019). Therefore, while the percentage of text similarities reported by TMS is an indication that academic misconduct may have occurred, it requires further investigation on the part of educators.

When discussing TMS, the internet-based similarity detection service marketed under the trade name *Turnitin* is often the first one that comes to mind. This American software was launched in 1998 by the company *iParadigms* and is currently used by over 16,000 universities, colleges, and high schools across several countries (www.turnitin.com). At its introduction, Australian and European universities, especially British institutions, embraced *Turnitin* as a useful tool to detect plagiarism. However, North American experts in the field of academic integrity expressed concerns about privacy issues, among other things, that will be discussed in the next section (Sutherland-Smith & Carr, 2005). One of *Turnitin*'s European competitors, called *Ouriginal* (a merger between *Urkund* and *PlagScan*), joined the "*Turnitin* family" in 2021. Although *Turnitin* supports more than 10 different languages, including English and French, there is another contender on the French market, named *Compilatio*, that was launched in 2005 and is distributed in 38 countries (www.compilatio.net). In addition to these large for-profit corporations there are some open-source options that may not be as sophisticated as *Turnitin* or *Compilatio* but are free of charge. For example, *Scribbr* appears at the top of a *Google* search for "plagiarism checkers," with *Grammarly* following not too far behind on an extensive list of alternatives.

Benefits and Limitations of TMS

On the one hand, TMS may be seen as time-savers for educators who wish to quickly identify evidence of plagiarism when marking student papers. According to Wheeler and Anderson (2010), not all educators fully understand what constitutes plagiarism or how to detect it. Therefore, it is useful to provide them with tools to assist in highlighting students' unoriginal work which may lack attributions or has been improperly cited. In that regard, TMS will produce detailed "originality reports" indicating a text-matching score, by categorizing excerpts into (1) properly quoted sources; and (2) potential problematic referencing (percentage of similarities to existing texts). The educator can then verify the original sources to confirm if a student indeed copied and pasted from a source without proper attribution. When educators deal with students who present limited knowledge and experience with bibliographic standards, TMS may be used as formative tools to help students improve their referencing skills (Limma et al., 2023). This can be done by way of exchanges between educators and students or simply by having students access their own text-matching reports. All in all, TMS can act as a plagiarism deterrent if students submit their assignments through such software (Bennett, 2005) or if they are aware that their teachers use TMS to conduct verifications (Davis, 2007).

On the other hand, TMS have been a source of controversy. First, they encourage a presumption of guilt which may impact negatively on the relationship between students and educators (or other institutional representatives) (Moore Howard, 2001). Over the years, there have been several litigation cases across Canada and the United States in which individuals or groups of students have refused to submit their assignments through TMS. Many of these disputes used arguments related to the violation of privacy and copyrights (Foster, 2002). For example, in 2006, the Senate of Mount Saint Vincent University (MSVU) in Nova Scotia (Canada) argued that students' academic work should not become part of a database accessed by external parties (MSVU, 2006). Nowadays, the positions adopted by academic institutions toward TMS vary from banning their usage to offering them as alternatives or making them compulsory. While discussions about access to students' personal data and breaches of intellectual property are still ongoing, there are other concerns that have been raised regarding TMS. There is a risk that "introducing students to [TMS] may simply teach them better strategies for avoiding detection" (Davis & Carroll, 2009, p. 61). For example, the software mainly recognizes similarities by comparing an assignment with texts written in the same language. Therefore, it is less likely that plagiarism would be detected if a student translated an excerpt for use in their assignment without attribution (Wager, 2014). Moreover, TMS focus on matching information in written format, discounting the use of other electronic sources by students to draft their assignments, such as audio and video files (Butakov & Scherbinin, 2009). Lastly, the advent of GenAI has significantly changed the landscape of TMS such that the capacity to recognize instances when writing styles may change within a given text, be it between human writers or machines generating human languages, must be developed.

TMS and GenAI

After the launch of *ChatGPT* by *OpenAI* in November 2022, an explosion of interest in GenAI tools was observed across all levels of education. Although this innovation was embraced by some educators, others expressed serious concerns with the potential for breaches of academic integrity that would go undetected (McMurtrie, 2023; Moloney et al., 2023). From a plagiarism point of view, the use of GenAI by students to draft their assignments (in part or as a whole) is different than borrowing the work of another author to pass it off as their own. At the present, when using GenAI, it is not possible to know the sources of information used by the machine to produce a text. Moreover, machines (including robots) are not currently recognized as authors, especially in academic realms. The main reason for this is because any attribution of authorship carries with it accountability for the work, and "as non-legal entities, [AI tools] cannot assert the presence or absence of conflicts of interest nor manage copyright and license agreements" (COPE, 2023). Therefore, there is a potential for new types of plagiarism different from the fundamental definition provided at the beginning of this chapter. The use of GenAI by students to generate text, without prior autorization by educators, is akin to fraudulent behaviours associated with cheating, such as when a student asks or pays someone else to draft their assignments (Eaton, 2021). The question that then needs to be asked is: If TMS are to be used to identify GenAI in students' assignments, what are educators trying to detect exactly?

Even though TMS were not designed to differentiate texts generated by machines from those drafted by humans, they were the first solution thought of to detect GenAI in student assignments. This was of particular concern for written assessments, such as essays and reports (McDermott, 2023), for which the use of GenAI makes it impossible to ensure the authenticity of submissions. Therefore, a myriad of tools was released during the year following the launch of ChatGPT, with claims that they were able to accurately diagnose whether texts were drafted by humans or GenAI. Among these tools was the response by *OpenAI* called *GPTZero*, as well as a plugin for *Turnitin*, and several other tools developed by independent programmers (Ali, 2023). For example, the Canadian company *Draft & Goal*, that was already working on a GenAI solution, released the beta version of a content detector in English and French (https://detector.dng.ai/). Most of these options promised to detect the patterns found in large language models (LLMs) able to generate texts following a training period. However, the development of detection capabilities proved to be challenging, especially in the face of "hallucinations," that is when GenAI systems "make up texts, answers or references in instances where they do not know how to reply" (Floridi, 2023, p. 3).

The poor reliability of these early detection tools resulted in students being accused of plagiarism based on inaccurate judgments whether assignments contained AI-generated text (Carter, 2023). In some cases, false-positive results were used to sanction students who then sued their institutions for impeding their academic journey and future job prospects (Merod, 2023; Mollenkamp, 2023). Given that these incidents made international headlines, many GenAI detectors were quickly taken off the market with commitments that they would only be released again after thorough investigation and improvement (Epstein-Gross, 2023). In Fall 2023, *Compilatio* launched its latest TMS version called *Magister+* claiming that it could accurately detect GenAI content in written assignments produced by students who used *ChatGPT* or one of its main competitors, such as *Bard* (the Google chatbot prototype) and *YouChat* (a chatbot linked to the new You.com search engine) (Perrier, 2023). In that regard, educators could view an independent line in similarity reports that provides a percentage of AI detection, in addition to the usual percentages related to text matching, text in quotes, and unrecognized language. At the time that this chapter is being written, the reliability of these reports has not been confirmed. Thus, it appears that the only way of detecting the unauthorized use of GenAI in a student's assignment would be if the student admitted to their wrongdoing or if the student was asked to draft a text under a controlled environment to compare with the assignment that is in question.

The Usage of TMS: Findings from an International Partnership

This paper draws on research by an international partnership on university plagiarism prevention involving at least 60 scholars from over 30 institutions across North America and Europe. The aim is to create new knowledge on competencies leveraged, learned, and taught during undergraduate programs to prevent plagiarism in written assignments. The research team developed data collection instruments based on a theoretical framework by Peters (2015), categorizing digital strategies into three types of skills: informational, writing, and referencing. During the seven-year (2021–2028) mixed methods study, there are three phases of data collection: (1) questionnaires, (2) individual and group interviews; and (3) recorded writing tasks. In this section, I focus on the questionnaires administered to university professors and undergraduate students.

During the period between February 15 and October 31, 2023, two separate questionnaires were respectively filled out by 1357 professors and 4664 students across a total of 28 universities, that are geographically categorized in Table 3.1 below.

The sociodemographic details for professors included gender identification (Female = 60%; Male = 39%; Others = 1%). These respondents were from three distinct regions: Canada (54%); Europe (35%); and the United States (11%). Their teaching languages were French (53%); English (36%); and Others (11%), among which 83% of respondents stated that they were teaching in their first language (mother tongue). The average number of years of teaching was 15.2 years, including 23% of respondents who had only cumulated between 1–5 years of experience and 117 respondents who had taught between 30–45 years. There is a difference in proportion of status between Lecturers (33%) and Professors (47%). In the latter category, only 229 out of 638 respondents identified themselves as Full Professors, whereas other respondents indicated that they were employed in Assistant Professor, Associate Professor or simply Professor positions.

Regarding sociodemographic details for students, the sample characteristics included gender identification (Female = 62%; Male = 34%; Others = 4%). These respondents were from three distinct regions: Europe (45%); Canada (42%); and the United States (13%). The main language used in their study programs were French (56%); English (32%); and Others (12%), among which 77% of respondents stated that they were studying in their first language (mother tongue). The average number of years spent in university as a student was 3.11 years, including 25% of

Table 3.1 Regional distribution of universities and respondents to questionnaires

	Canada	United States	Europe	Sub-totals
Universities	13	7	8	28
Professors	736	148	473	1357
Students	1949	601	2114	4664
Total (respondents)	2685	749	2587	6021

Table 3.2 Professors-Q.45: I use text-matching software (*Turnitin*, *Urkund*, *Compilatio*) to help me identify plagiarism in student work (frequency scale)

Never	Rarely	Sometimes	Frequently	Almost always or always	No answer
450	88	136	109	233	39
42.7%	8.3%	12.9%	10.3%	22.1%	3.7%

Table 3.3 Students-Q.66: When I think that my teachers use text-matching software (*Turnitin*, *Urkund*, *Compilatio*, etc.) I am more careful not to plagiarize when writing my assignments (agreement scale)

Strongly disagree	Disagree	Neither agree nor disagree	Agree	Strongly agree	No answer
453	498	697	1091	957	312
11.3%	12.4%	17.4%	27.2%	23.9%	7.8%

Table 3.4 Students-Q.67: I use text-matching software to help me identify plagiarism in my own assignments (*Turnitin*, *Urkund*, *Compilatio*, etc.) (frequency scale)

Never	Rarely	Sometimes	Frequently	Almost always or always	No answer
2249	523	602	326	234	74
56.1%	13.1%	15.0%	8.1%	5.8%	1.9%

respondents who were in their first year of university studies and 84 respondents who had been studying at university for over 10 years. Among all students surveyed, 29% were attending university after an average gap of 6.2 years in education.

The questionnaires, administered in English and French, included a total of 49 questions for professors and 68 questions for students. The two instruments were divided into six sections: (1) sociodemographic; (2) general information; (3) informational skills; (4) writing skills; (5) referencing skills; and (6) plagiarism knowledge. In the last part of both questionnaires, questions about text-matching software were included in the form of statements rated by respondents on 5-point frequency or agreement Likert scales. The respondents also had the option of selecting "no answer" if they did not understand the question, felt that it did not apply to their situation or preferred not to answer. The answers to the three questions posed are provided in Tables 3.2, 3.3, and 3.4 below, as descriptive quantitative results.

According to the findings, the majority of university professors (42.7%) who responded to our questionnaire "never" use TMS to identify plagiarism in student work. However, when breaking down these results by regions, we find that 83% of those respondents were from Canadian universities. Besides the litigation at Mount Saint Vincent University that was previously mentioned, other disputes originating from Canadian universities resulted in several institutions either banning TMS or rendering them optional. This may have influenced their usage by respondents of that region, who are also coming from 15 Canadian universities, among which eight are francophone. These French-speaking institutions may be reluctant to purchase licenses for an American software (*Turnitin*) that is dominating the market or a

French software (*Compilatio*) that has not been on the market for a long time. Regarding the professors (22.1%) who "almost always or always" use TMS, 76% of those respondents were from European universities, including British institutions that have been proponents of TMS since *Turnitin* was introduced in the late 1990s. Based on our data, it is not possible to know if there was an increase in the usage of TMS by university professors, over the past year, due to the rapid growth of GenAI tools.

The findings also reveal that undergraduate students "agree" (27.2%) or "strongly agree" (23.9%) that they are more careful not to plagiarize when they think that their professors use TMS. The formulation of the questionnaire statement using the verb "think", instead of "know" leaves room for uncertainty, but it seems to be enough to motivate students to avoid plagiarizing. As for the usage of TMS by students, the majority of respondents (56.1%) answered that they "never" use such a tool to help them identify plagiarism in their own assignments. To do that, their professors would have had to give them access to reports produced by the software. Moreover, they would need to understand how to interpret such reports, which may require specific software familiarization training. In that respect, one of the limitations of our questionnaire would be the understanding of what TMS constitute when responding to the questions on that topic. For example, this may be the reason why 25.2% of the respondents answered "neither agree nor disagree" or "no answer" to the question regarding the use of TMS by their professors.

There are two other limitations regarding this study, including a disproportionate number of students (30.8%) from one of the European universities who responded to the questionnaire, which is influencing the interpretation of data from that region and the comparison of regional dataset. Finally, the length of both questionnaires seems to be problematic when realizing that 41.2% of professors and 44.0% of students who started answering sociodemographic questions did not complete the questionnaire. Therefore, when condensing data, we decided to include incomplete questionnaires showing data captured in the first three sections (sociodemographic, general information, informational skills) but to exclude all other unfinished questionnaires (n = 436 for professors and n = 2494 for students). Given the volume of data collected, the impact of these exclusions on research findings is considered to be limited.

Finally, our students' questionnaire also included the following question: "To help me with my written assignments, I use artificial intelligence tools (ChatGPT, AI-Writer, QuillBot, etc.)." Therefore, we completed a correlation analysis with the *SPSS* software to find out if students using TMS were more likely to also use AI tools. The results obtained demonstrated *odds ratio* of 2.4 that were statistically significant with a *p value* of 0, meaning that there is a strong association between the use of GenAI tools and TMS, as well as a low likelihood that this association is the result of chance. These results could simply be a reflection of those students who are comfortable with technology and wanting to maximize the use of any digital tools in their learning process to better their chances of success. Otherwise, some students may be using both GenAI and TMS to plagiarize in their assignments by having machines produce their papers and exploiting reports to adjust the generated text just enough to bring similarity scores down. In any case, when it comes to plagiarism, educators should move away from detection and sanction to adopt a preventative approach to promote a culture of academic integrity.

The Focus on Plagiarism Prevention with Text-Matching Software

In the previous two sections of this chapter, I have established that a large proportion of educators and students are using TMS during evaluation processes, even though TMS are not currently able to detect AI-generated text. In keeping with the evolution of teaching and learning practices, I next discuss how TMS can be used to provide feedback (by educators) and to self-assess (by students) with a view to preventing plagiarism.

The Use of TMS by Educators to Provide Constructive Feedback

According to Hayden et al. (2021), TMS represent "one tool used as part of a comprehensive instructional and institutional academic integrity strategy" (p. 9). The reports obtained from these software provide valuable information to better understand the skills that students either master or require to improve. The "over-reliance on some sources, incorrect or incomplete citation, and insufficient paraphrasing" (Davis & Carroll, 2009, p. 58) are among writing issues to be discussed with students based on information drawn from TMS. As such, educators can make use of similarity reports to provide dedicated support to individual students who may be falling behind the rest of the group for various reasons, including studying in a second language (Sutherland-Smith, 2008) or returning to university after gap years. Furthermore, educators may choose to use TMS reports to provide feedback to the whole class, if it is considered that the collective would benefit from a levelling of individual skills (informational, writing, or referencing). If educators lack the resources or knowledge to develop students' skills to prevent plagiarism, it is possible to invite a librarian for a workshop or refer students to other support services provided by the institution, such as peer mentors or tutors.

In addition, students should be taught how to properly use GenAI to produce their assignments because universities have the responsibility to prepare students for the labour market where AI will be overwhelmingly present in the future. To do this, educators need to be crystal clear in their expectations regarding the specific tasks that can be accomplished with AI. Over the past year, I have trialled various ways of integrating GenAI in my classes, including allowing students to use ChatGPT to assist them in drafting a literature review. Those wishing to use GenAI to produce their paper had to submit a declaration of academic integrity stating which AI tool was used and how. Even though I had taught all students how to formulate effective prompts and follow on messages to communicate with ChatGPT, only a few decided to use it to complete various tasks, such as developing a plan, drafting an introduction, or improving transitions between paragraphs. The students' assignments were uploaded into the *Magister+* version of *Compilatio* and the software detected AI in every single document, with a warning that similarity scores were just indicators, not evidence of plagiarism, and that a verification of knowledge should be done with the authors. The excerpts highlighted as having potentially been generated by AI were discussed with the students. It became clear that most detected sections were not AI-generated and that parts of the assignments that were generated by AI were not flagged. This trial confirmed that TMS require further reprogramming to be recognized as reliable tools for detecting GenAI texts.

The Use of TMS by Students for Self-Assessment and to Improve Skills

Given that some TMS offer the option to grant access to students to view their own report, they can use the information as an interim step to address any issues prior to the final submission of their assignments. This is the case for the *Turnitin* family of software, but not for *Compilatio*, even though the latter markets its product as a pedagogical resource to increase awareness about authenticity. According to Bretag (2016), "if TMS are to be used, students should have the opportunity to use them as an educative drafting tool to develop their writing and referencing skills" (p. 30). Even though Peacock et al. (2006) favour the use of TMS through tutor guidance, Davis (2007) states that most students wish to have their own access to view reports on draft assignments.

If the latter avenue is considered, then training must be given to students, so they can develop sufficient understanding of conventions to interpret similarity reports. However, knowing how to read reports is not sufficient. Students must also know how to problem-solve to address a myriad of academic writing issues related to citing and paraphrasing (Roussey & Piolat, 2005). An additional feature available in some TMS is a peer review option that could be done between students as a formative assessment. In any case, if students have access to their similarity reports, they should not be allowed to submit each assignment more than twice, to avoid changes made simply to reduce similarity scores, instead of actually learning to improve academic literacy.

Conclusion

The recent fast-paced evolution of GenAI has certainly changed the way in which plagiarism is defined, detected, and prevented. This chapter established the advantages and disadvantages of TMS, before and after the launch of ChatGPT. In this regard, it was determined that TMS, in their current form, are not able to reliably detect text generated by machines. Moreover, the findings from an international research partnership on plagiarism prevention revealed that a majority of university professors and undergraduate students use TMS during the evaluation processes of written assignments with or without the assistance of GenAI. As the GenAI landscape continues to evolve at a fast pace, it will be interesting to compare our preliminary findings to future studies, especially when TMS become more reliable over time. Finally, teaching strategies were recommended for educators to show students how to integrate GenAI in written documents, as well as to leverage TMS reports to provide individual or group feedback and enable students to use that same information for self-assessment. These constructive pedagogical approaches will contribute to the development of useful skills to prevent plagiarism in institutions of higher education and instill ethical practices in future citizens.

References

Ali, F. (2023, January 6). GPTZero set to counter AI plagiarism caused by services like ChatGPT. *Neowin*. https://www.neowin.net/news/gptzero-set-to-counter-ai-plagiarism-caused-by-services-like-chatgpt/

Bearman, M., & Luckin, R. (2020). Preparing university assessment for a world with AI: Tasks for human intelligence. In M. Bearman, P. Dawson, R. Ajjawi, J. Tai, & D. Boud (Eds.), *Re-imagining university assessment in a digital world* (pp. 49–63). Springer. https://doi.org/10.1007/978-3-030-41956-1

Bennett, R. (2005). Factors associated with student plagiarism in a post-1992 university. *Assessment and Evaluation in Higher Education, 30*(2), 137–162.

Bretag, T. (2016). Educational integrity in Australia. In T. Bretag (Ed.), *Handbook of academic integrity* (pp. 23–38). Springer.

Butakov, S., & Scherbinin, V. (2009). The toolbox for local and global plagiarism detection. *Computers & Education, 52*(4), 781–788. https://doi.org/10.1016/j.compedu.2008.12.001

Carter, T. (2023, September 22). Some universities are ditching AI detection software amid fears students could be falsely accused of cheating by using ChatGPT. *Business Insider*. https://www.businessinsider.com/universities-ditch-ai-detectors-over-fears-students-falsely-accused-cheating-2023-9

Committee on Publication Ethics (COPE). (2023). *Authorship and AI tools*. https://publicationethics.org/cope-position-statements/ai-author

Davis, M. (2007). The role of Turnitin in the formative process of academic writing: A tool for learning and unlearning? *Brookes e-Journal of Learning and Teaching, 2*(1), 1–9.

Davis, M., & Carroll, J. (2009). Formative feedback within plagiarism education: Is there a role for text-matching software? *International Journal for Educational Integrity, 5*(2), 58–70. https://doi.org/10.21913/IJEI.v5i2.614

Eaton, S. E. (2021). *Plagiarism in higher education: Tackling touch topics in academic integrity*. Libraries Unlimited.
Epstein-Gross, C. (2023, July 26). OpenAI abruptly shuts down ChatGPT plagiarism detector- and educators are worried. *Observer*. https://observer.com/2023/07/openai-shut-ai-classifier/
Flach, P. (2012). *Machine learning: The art and science of algorithms that make sense of data*. Cambridge University Press. https://doi.org/10.1017/CBO9780511973000
Floridi, L. (2023). AI as agency without intelligence: On ChatGPT, large language models, and other generative models. *Philosophy & Technology, 36*(15), 1–7. https://doi.org/10.1007/s13347-023-00621-y
Foster, A. L. (2002, May 17). Plagiarism-detection tool creates legal quandary. *The Chronicle of Higher Education*. https://www.chronicle.com/article/plagiarism-detection-tool-creates-legal-quandary/
Hayden, K. A., Eaton, S. E., Pethrick, H., Crossman, K., Lenart, B. A., & Penaluna, L.-A. (2021). A scoping review of text-matching software used for student academic integrity in higher education. *Education Research International*, 1–15. https://doi.org/10.1155/2021/4834860
Jamieson, S., & Moore Howard, R. (2019). Rethinking the relationship between plagiarism and academic integrity. *International Journal of Technologies in Higher Education, 16*(2), 69–85.
Jeffrey, D., & Dias, W. (2019). Perceptions du plagiat par de futurs enseignants québécois: dimensions éthiques et pédagogiques. *Canadian Journal of Education, 42*(3), 767–790. https://www.jstor.org/stable/26891584
Limma, P., Kraiwanit, T., Jangjarat, K., Klayklung, P., & Chocksathaporn, P. (2023). The use of ChatGPT in the digital era: Perspective on chatbot implementation. *Journal of Applied Learning & Teaching, 6*(1), 1–10. https://doi.org/10.37074/jalt.2023.6.1.32
McDermott, M. (2023, March 14). Does ChatGPT pose a threat to the future of the college essay? *Utah Public Radio*. https://www.upr.org/utah-news/2023-03-14/does-chatgpt-pose-a-threat-to-the-future-of-the-college-essay
McMurtrie, B. (2023, March 6). ChatGPT is everywhere. Love it or hate it, academics can't ignore the already pervasive technology. *The Chronicle of Higher Education*. https://www.chronicle.com/article/chatgpt-is-already-upending-campus-practices-colleges-are-rushing-to-respond
Merod, A. (2023, June 7). Turnitin admits there are some cases of higher false positives in AI writing detection tool. *K-12 Dive*. https://www.k12dive.com/news/turnitin-false-positives-AI-detector/652221/
Mollenkamp, D. (2023, July 11). How does a tool that detects cheating with ChatGPT grapple with 'false positives'? *EdSurge*. https://www.edsurge.com/news/2023-07-11-how-does-a-tool-that-detects-cheating-with-chatgpt-grapple-with-false-positives
Moloney, A., Baptista, D., & Teixeira, F. (2023, July 5). Despite cheating fears, Latin American universities embrace ChatGPT. *New Straits Times*. https://www.nst.com.my/opinion/columnists/2023/07/927852/despite-cheating-fears-latin-american-universities-embrace-chatgpt
Moore Howard, R. (2001, November 16). Forget about policing plagiarism: Just teach! *The Chronicle of Higher Education*. https://www.chronicle.com/article/forget-about-policing-plagiarism-just-teach/
Mount Saint Vincent University. (2006). *Minutes of senate meeting*. https://www.msvu.ca/wp-content/uploads/2020/05/MinutesMar606.pdf
Okerlund, J., Klasky, E., Middha, A., Kim, S., Rosenfeld, H., Kleinman, M., & Parthasarathy, S. (2022). *What's in the chatterbox? Large language models, why they matter, and what we should do about them* (Technology Assessment Project Report). University of Michigan.
Peacock, S., Sharp, J., & Anderson, S. (2006). Pickup a plagiarism practice? A holistic approach to deploying Turnitin UK as a sustainable driver in altering institutional plagiarism practice. In *Proceedings of the JISC International Plagiarism Conference, Newcastle, UK*.
Perrier, M.-C. (2023, March 14). ChatGPT: le spécialiste de l'anti-plagiat prépare un logiciel pour détecter les devoirs écrits grâce à l'intelligence artificielle. *FranceInfo*. https://france3-regions.francetvinfo.fr/auvergne-rhone-alpes/haute-savoie/annecy/chatgpt-le-specialiste-de-l-anti-

plagiat-prepare-un-logiciel-pour-detecter-les-devoirs-ecrits-grace-a-l-intelligence-artificielle-2731862.html

Peters, M. (2015). Enseigner les strategies de ceréacollage numérique pour éviter le plagiat au secondaire. *Canadian Journal of Education, 38*(3), 1–28. https://journals.sfu.ca/cje/index.php/cje-rce/article/view/1878

Roussey, J.-Y., & Piolat, A. (2005). La révision du texte: une activité de contrôle et de réflexion. *Psychologie Française, 50*, 351–372. https://doi.org/10.1016/j.psfr.2005.05.001

Sutherland-Smith, W. (2008). *Plagiarism, the internet and student learning: Improving academic integrity*. Routledge.

Sutherland-Smith, W., & Carr, R. (2005). Turnitin.com: Teachers' perspectives of anti-plagiarism software in raising issues of educational integrity. *Journal of University Teaching and Learning Practice, 2*(3), 94–101.

Wager, E. (2014). Defining and responding to plagiarism. *Learned Publishing, 27*, 33–42. https://doi.org/10.1087/20140105

Wheeler, D., & Anderson, D. (2010). Dealing with plagiarism in a complex information society. *Education, Business and Society: Contemporary Middle Eastern Issues, 3*(3), 166–177. https://doi.org/10.1108/17537981011070082

Chapter 4
'Releasing Something Dangerous into the Wild': A Media Ecological Perspective on the GenAI—Academic Integrity Nexus

Felix Odartey-Wellington and Sarah MacRae

Abstract A May 2023 article in *The New York Times* reported the alarm of industry insiders that the continued advancement of GenAI technologies has the unintended consequence of "releasing something dangerous into the wild." This ecological analogy is our point of departure. Our chapter approaches the GenAI-Academic Integrity nexus from a media ecological standpoint. Nested in critical communication studies, this perspective assumes complementarities between communication technologies and societal change. In the context of this book, to what extent does GenAI problematize academic gate-keeping, and what do the societal changes attendant to the existence of GenAI mean for academic citizenship? We draw from our observations teaching communication studies in a small university in Atlantic Canada which has situated the GenAI-Academic Integrity question within the context of academic citizenship, and is evolving policies and best practices to address GenAI use within the academy.

Keywords GenAI · Cape Breton University (CBU) · Media ecology · Academic citizenship · Academic integrity

Introduction

As technology that imitates autonomous human cognitive functions, Generative Artificial Intelligence (GenAI) is topical, especially the ramifications for academia of applications like ChatGPT that "democratize access to Large Language Models" (García-Peñalvo & Vázquez-Ingelmo, 2023, p. 1) or AI technology capable of

F. Odartey-Wellington (✉) · S. MacRae
Cape Breton University, Sydney, NS, Canada
e-mail: felix_odartey@cbu.ca; sarah_macrae@cbu.ca

A. E. King (ed.), *Artificial Intelligence, Pedagogy and Academic Integrity*, Advances in Artificial Intelligence in Education 1,
https://doi.org/10.1007/978-3-031-92534-4_4

originating content. A May 2023 article in *The New York Times* reported the alarm of industry insiders that the continued advancement of AI technologies has the unintended consequence of "releasing something dangerous into the wild" (Metz, 2023). This ecological analogy is our point of departure. Our chapter approaches the GenAI-Academic Integrity nexus from a media ecological standpoint. Nested in critical communication studies, this perspective assumes complementarities between communication technologies (media) and societal change. In the context of this chapter, to what extent does the "migration" or "invasion" of GenAI into the academic ecosystem problematize academic gate-keeping and what do the societal changes attendant to the existence of GenAI mean for academic citizenship? We draw from our observations in a small university in Atlantic Canada which is evolving policies and best practices to address GenAI use within the academy. Methodologically, this is therefore, a case study that allows for the sharing of current challenges and approaches with respect to the subject of GenAI and academic integrity.

The complementarity between media evolution and academic integrity has animated academic discourse in Canada, and Sarah Eaton and Julia Christensen Hughes have edited a comprehensive body of work on the subject (Eaton & Christensen Hughes, 2022a). They, however, acknowledge the need to further account for perspectives from Atlantic Canada and northern Canadian territories (Eaton & Christensen Hughes, 2022c). This chapter is intended to fill this gap as regards Atlantic Canada. As well, the authors of this chapter teach general writing and communication courses, which are identified as having a high potential for academic cheating (Crossman, 2022).

GenAI, Academic Integrity and Media Ecology

When one thinks of academic integrity, what jumps to mind is plagiarism, though academic integrity covers more than that, especially when situated in the context of academic citizenship. However, plagiarism is a good place to start in a communication studies-focused approach to the GenAI-Academic Integrity nexus. Plagiarism or some form of intellectual property appropriation is not new. Indeed, a Canadian lawyer once argued on behalf of their client accused of operating an essay-writing mill that professional essay-writing shares a place with sex work as one of the world's oldest professions (Murray, 1989, p. A7). As pointed out by Geoff Buerger in his detailed history of the evolution of academic cheating, "[t]he issues facing educators in the day of the Internet remain much the same as those with which they have dealt at least as far back as the Great Depression" (2021, p. 3).

Media ecology as a school of thought informing contemporary communication studies emerged in twentieth century North America as electronic media proliferated (Strate, 2004). As can be seen in Strate's (2004) extensive review of the literature, scholars like Neil Postman, Joshua Meyrowitz, Harold Innis, Marshall McLuhan, Walter Ong, Lewis Mumford, and Elizabeth Eisenstein have influenced this theoretical formation. As previously indicated, media ecology or "medium theory" (Meyrowitz, 1985) considers complementarities between media evolution and societal change. As Postman puts it: "Technological change is neither additive nor subtractive. It is ... 'ecological' in the same sense as the word is used by environmental scientists. One significant change generates total change" (1992, p. 18).

In this vein, if language and words are considered to be media (Mumford, 1934, 1967; Ong, 1982; Postman, 1992), the potential for plagiarism existed the very moment society had not just words but a concept of communication and intellectual property ownership. Therefore, a 40 AD Roman poet is credited with originating the term "plagiarism" when he lampooned rivals who were using his lines without attribution (Algra, 2021). As the advent of the printing press gave rise to mass copying and mass communication, it is no surprise that the period between the fifteenth and seventeenth centuries is seen as the locus of the concept entering English language usage (Algra, 2021). Technological shifts in communication production gave rise not just to the commodification of communication through print capitalism but also a growth in the concept of ownership of ideas in a normative sense (Okerson, 1991).

Key to the media ecological perspective is that there are pros and cons attendant to the invention of any medium, and hence, a need to critically evaluate media evolution. In this regard, the emergence of a new medium potentially disrupts the dominance of older media and societal power relations. For instance, the emergence of the printing press democratized access to literacy skills and disrupted scribal and ecclesiastical power (Thompson, 1995). Another significant point is one of unintended consequences (Postman, 1992). For example, it is pointed out that: "those who had worked on the development of the technology that evolved into the Internet had never anticipated that those who used it might do so for nefarious reasons" (Eaton & Christensen Hughes, 2022b, p. 11). The introduction of every new medium, therefore, is the potential release of something dangerous into the societal ecosystem. However, media ecology provokes an understanding that concerns about contemporary media impacts might not necessarily be novel (Ong, 1982; Standage, 2018), and it is possible to place new media within prior-existing normative frameworks, with necessary modifications.

The foregoing is relevant to the GenAI-academic integrity nexus because instructors have historically enjoyed the power of academic gate-keeping or quality control, dependent, in part, on the ability to detect academic cheating such as plagiarism. Prior to the emergence of GenAI, students had access to means of plagiarism and other forms of cheating (Walsh, 2025). Indeed, years ago, French philosopher Jean Baudrillard argued that in this electronic age, media attenuate the dichotomy between the original and the copy or "simulacrum" (Baudrillard, 1988, p. 6). While the availability of the Internet and other forms of digital technology create fantastic opportunities for scholarship and academic productivity, they also have the unintended consequence of creating opportunities for plagiarism and other forms of cheating in ways that did not exist before. These developments have challenged the academic gate-keeping power needed to ensure the integrity of higher education. For instructors,

> [t]he continued advancement of technology-assisted educational practices has contributed to the continuing rise of academic misconduct and while software for detecting plagiarism can help ease the burden of verifying referenced material, it is costly and not available at all educational institutions. This has left faculty feeling frustrated and discouraged, as they attempt to stay one step ahead of new and innovative cheating methods. (Hamilton & Wolsky, 2022, p. 469)

The same sentiments are expressed today in response to GenAI academic misconduct (Walsh, 2025). Academia has also evolved normative frameworks to account for new forms of digital media-driven academic misconduct. For example, it is noted that the founding of the International Center for Academic Integrity in 1992 and the formulation of its widely-adopted normative values coincided with the popularization of the Internet in the early 1990s (Eaton & Christensen Hughes, 2022b). However, beyond the Internet, the emergence of GenAI has further problematized academic gate-keeping power, no doubt a source of alarm within academia (Walsh, 2025). As has been observed, the advent of ChatGPT (or Chat Generative Pre-Trained Transformer) in 2022 broke new ground in the digital media ecosystem and academia: "By substantially lowering the barrier for individuals without a technological background to utilize the generative capabilities of LLMs , the advancements of ChatGPT and similar models have propelled generative AI (GenAI) into the forefront of educational discourse, igniting a mixed response of excitement and concern" (Deng et al., 2024, pp. 1–2). Therefore, while GenAI potentially expands the opportunities for genuine academic productivity, the technology is potentially a dangerous species in the academic ecosystem. At the same time, it is important to recognize that this is not particularly novel: when crowdsourced knowledge platforms like Wikipedia emerged, educators were similarly concerned about implications for good scholarship. As it evolved, however, Wikipedia became recognized for its research, teaching and learning potential, with the appropriate caveats (Zhao & Bishop, 2011). In that vein, Verrax and Scheid write that: "While there are fears that generative AI poses civilisational risks and obliges us to entirely rethink education, some perceive it as yet a breakthrough innovation similar to the printing press, radio, computers, and the Internet, none of which led to the disappearance of schools or universities" (2024). Ong also reminds us that: "Most persons are surprised, and many distressed, to learn that essentially the same objections commonly urged today against computers were urged by Plato in the Phaedrus and in the Seventh Letter against writing" (1982, p. 79).

GenAI in the Classroom: A CBU Experience

As Martine Peters and Catherine Déri discuss in this volume, the introduction of GenAI has problematized academic integrity regimes. It bears mentioning that McLuhan co-authored an article in the 1960s predicting that the literacy proficiency dominance which instructors enjoyed vis-à-vis students in the traditional literacy cultures undergirding modern academia (Innis, 1951; Eisenstein, 1983) would be attenuated in the electronic age (McLuhan & Leonard, 1967). The advent of digital media and GenAI as an offshoot, has fulfilled this prophecy and upended this dominance, as can be gleaned from other chapters in this volume which illustrate that when it comes to GenAI, adoption (and proficiency) among students is often higher than what obtains with instructors. Academic institutions and the instructors they employ have been compelled to play catch-up (Cottom, 2025; Walsh, 2025). Therefore, when it comes to the threat that GenAI poses to academic integrity, Cape Breton University (CBU) is no different from colleges and universities across the world. A primarily undergraduate university located on an island off mainland Nova Scotia in Atlantic Canada, CBU has seen an exponential growth in its student population from under 3000 in 2017 (Armstrong, 2022) to more than 9000 in 2023 (with about 77% being international students) (Ayers, 2023), with students enrolled in online, on-campus and Indigenous community-based courses. A university that used to boast small class sizes now has to contend with large classes in some of its programs, to the extent that it utilizes space in a local movie theatre for some lectures (Choi & Hou, 2023). This has coincided with the rise of GenAI applications and their use by students. As instructors in the communication program, we are able to share observations regarding the use of GenAI in the classroom.

Canadian colleges and universities experienced an upsurge in academic cheating in the 2020–2022 timeframe when, due to COVID-19 pandemic public health measures, most courses had to be offered online (Choi & Hou, 2023). CBU was no different, and it is within this context that the advent of GenAI use in the institution is situated. From the tactical level as communication instructors, we can say that student use of GenAI today occurs in three ways:

1. The use of GenAI text rewriting programs;
2. The use of GenAI to answer assignment questions or complete coursework, such as tests and term papers; and,
3. The use of GenAI in controlled environments, such as instructor-led explorations of how AI can be used responsibly.

In the first instance, we have observed an uptick in students relying on GenAI text rewriting software to render their work in what non-native English speaking students believe is more presentable "Canadian English" or to defeat plagiarism detection. However, while what the software produces is often unintelligible, as Juola illustrates in this volume, unless the user has competent English language skills, they might not detect this. As well, the technology would not help a student articulate their own voice, enact their agency, or develop strong literacy, communication and critical thinking skills (Deng et al., 2024). In the end, students sacrifice process for product, to their intellectual detriment. (As Deng et al. put it: a prioritization of "convenience over engagement with challenging learning materials and projects" (2024, p. 31)). Thus, even when the use of GenAI results in superlative course work that suggests student growth, "[a]n alternative explanation for the improved academic performance could be the higher quality of work produced with ChatGPT's assistance, which might be misconstrued as genuine improvement in student learning" (Deng et al., 2024, p. 33).

Instructor response to the misuse of text rewriting programs is driven by where it falls within CBU's Academic Integrity policy. Where the intention is not to mislead but rather to "polish" writing, the approach is to provide students with more guidance, for example in terms of leveraging the institution's Writing Centre capabilities. Where the intent is to mislead (e.g., defeat plagiarism detection software), the approach is to apply corrective measures (e.g., additional assignments) and reinforce exhortation on appropriate citation styles and integrity matters. If the student does not demonstrate insight into their behaviour and persists, then other sanctions are applied as per the Academic Integrity policy.

In the instance where a student uses GenAI programs to answer test questions or write term papers, it is treated more seriously, with a potential for sanctions to be part of the student's record. However, in all of the above instances, the approach has been proactive exhortation to pre-empt breaches, with varying degrees of effectiveness in influencing student behaviour. We have found that the temptation to have a tool complete research and writing necessary for academic capital can be as irresistible as having power tools or heavy equipment available capable of completing what would ordinarily require physical effort.

Similar to other researchers (e.g., Lye & Lim, 2024; Tan et al., 2024), we have found that GenAI detection capabilities in plagiarism detection software are unreliable. There is, for example, the potential for false positives (and negatives) as other authors discuss in this volume, and thus, inaccurate assessment of work as AI-generated. Research shows that GenAI detection programs can be biased against non-native English speakers (Liang et al., 2023), which would be problematic in a university like CBU with a predominance of non-native English-speaking students. Where possible, the old-fashioned approach of meeting with a student to have them review their work and confirm authorship continues to be a viable approach to resolving issues around problematic course work submissions. This can confirm whether, for example, the student is able to read, explain or reproduce what they have submitted. In fact, we have also found that technological interfaces between students and teachers can problematize relationships vis-à-vis the GenAI question. A comment by a CBU professor to the media in this regard is revealing: "Because I know the voice of my students, I know how they speak, all of their idiosyncrasies and quirks. The second they switch to ChatGPT, it's like they just randomly adopted a British accent" (Moore, 2023). Therefore, the personal or human touch is key. For example, CBU's communication program prioritizes extemporaneous oral presentations in our courses. Orality is the oldest medium of communication (Ong, 1982; Strate, 2008) and its scholars laid the foundation for assessing authentic communication (Rolls, 2006). Hence, where courses require extemporaneous (rather than scripted or memorized) presentations, there are opportunities to assess students' authentic competencies. A similar recommendation has been recently made in the *Canadian Army Journal* for *viva voce* examinations to be integrated into professional military training in Canada in response to the GenAI issue (Richards, 2024). While opportunities to develop rapport with students is challenged by growing class sizes, we are still fortunate to have relatively smaller class sizes at CBU and particularly in our introductory communication courses, which include mandatory lab components where students work in small groups. Under these circumstances, it is possible to detect GenAI use. Timely meetings with students to review submissions and confirm authorship are also relatively possible to arrange. It is more challenging for colleagues in programs with large class sizes. As well, whereas academic cheating used to be confined to the margins, with the advent of GenAI, misuse in the first two instances provided in this paper sometimes accounts for the critical mass in a course rather than the outliers. Indeed, we have found a predominance of Gen-AI misuse in the writing of lab journals that require reflective writing which engages students at cognitive, affective and behavioural levels. In response, we have brought back the model of having students hand-write journals in the labs, as the dominant digital technology has not consigned previous writing technologies to obsolescence, even in the workplace for which students are being prepared. Students are, therefore, able to produce authentic work while exercising traditional literacy skills, in parallel with the use of digital literacy skills in other course contexts.

Our Department also employs media literacy to help students understand not just why unethical GenAI use is problematic but also, ways in which GenAI can be used responsibly. We are reminded that, even before today's digital age media ecosystem came into being, McLuhan projected that: "Education will become recognized as civil defense against media fallout. The only medium for which our education now offers some civil defense is the print medium. The educational establishment, founded on print, does not yet admit any other responsibilities" (McLuhan, 1964, p. 305). In other words, relevant media literacy must be an important component in contemporary teaching and learning. This has required instructors to familiarize themselves with AI evolutionary trends through peer learning (for example, we have benefitted from the work of a colleague who is the pioneering CBU Chair in Flexible Digital Delivery, as well as insight from work done by our colleagues in the English program) and regular programs organized by the university's Centre for Teaching and Learning. Consistent with GenAI pedagogical best practices (Tan et al., 2024), instructors therefore, integrate the subject of GenAI into classroom learning. This is the third instance in which AI use occurs within our context. For example, in our media ecology-informed courses at CBU, we demystify AI itself—after all it is in quotidian use in applications such as word processing software, virtual meeting platforms and search engines (Wakefield, 2023)—while creating awareness of its limitations. Indeed, as Déri observes in this volume, since GenAI has become part of the media ecosystem, students need the relevant exposure for success outside the classroom. As an illustration, in a communication research tools activity, a colleague required students to compare and contrast the CBU Library website, Google Scholar and ChatGPT as research tools. She noted: "Students at the end of the exercise said it was easiest to find sources on Google Scholar and found that ChatGPT helped them in brainstorming ideas, but they had to be actively engaged in the process/fact check the responses" (J. Whitehead, personal communication, 2023, October 27). Similarly, to explore Gen AI prompt literacy, as well as potential GenAI biases and limitations, one of the authors demonstrated through a workshop in their Development Communication class how GenAI "hallucinated" or completely made up false citations. Our focus, therefore, has been on what GenAI can contribute to academic work, and the need for due diligence when using these technologies. This is because, apart from ethical concerns, recent research suggests that "students often exhibit trust in the capacity of ChatGPT to deliver accurate answers" (Deng et al., 2024, p. 3).

Resonating with the observations in the foregoing, it pays to remember that the title of one of Postman's cautionary texts is aptly *Technopoly: The Surrender of Culture to Technology* (Postman, 1992). From a media ecological perspective, he bemoans continuous shifts from tool-using society to technocracy and then finally, to technopoly or "totalitarian technocracy" (Postman, 1992, p. 48) whereby tools are no longer subservient to humans but rather, attenuate human autonomy and agency. Even outside the GenAI context, we have observed instances where students uploade courseware and assignments to transactional "student help" sites and inadvertently reveal their (and sometimes other students') sensitive personal information in metadata or breach the copyright of courseware developers. We have observed cases where students submit papers with telltale GenAI boilerplate messaging (for example requests for a GenAI app to be rated for its generated content) that was not deleted.

Based on our experiences in the classroom and current affairs, we believe, therefore, that rather than simply focusing on whether AI will take over jobs in the future (Wakefield, 2023), another salient question is the extent to which humans can resist the temptation to surrender their agency to, or put themselves at the mercy of GenAI. We say this because of the increasing cases of humans abdicating responsibilities such as fact-checking, wordsmithing and proofreading to AI.[1] An example is a student submitting work with the following entry: "*As an AI language model, I don't have personal experiences, but I can provide hypothetical examples...*" Yet another example is a student submitting the following in their work: "*Since I'm a machine learning model, I don't possess personal emotions, experiences, or feelings. However, based on numerous accounts and examples I've been trained on, here's a hypothetical negative experience...*"

To sum up the view from the tactical level, we can say that GenAI misuse problematizes instructional gate-keeping capabilities particularly in terms of authenticity or integrity of academic work, and it has become important for instructors to be familiar with GenAI technology and relevant institutional policies to play their traditional roles as academic mentors and gatekeepers. Despite challenges, opportunities to directly interact with students and media literacy remain useful approaches to the GenAI-academic integrity question. However, instructor time is increasingly consumed by cases of GenAI misuse, which could be unmanageable if one had a rather large sample size of such cases.

[1]Valdez, J. (2023, October 19). Pras blames AI for his guilty verdict in conspiracy trial. Does he have the right culprit?; Milmo, D. (2023, June 23). Two US lawyers fined for submitting fake court citations from ChatGPT.

CBU's Approach to the GenAI - Academic Integrity Question

Thus far, we have talked about GenAI in the CBU classroom from the tactical level. At the strategic-operational level, CBU's response to the GenAI-academic integrity question is media ecology-driven. It starts with the understanding that the risks posed by GenAI are not necessarily novel; they have existed since mediatization of some sort became a thing. To that extent, there are prior-existing norms that account for AI-driven academic misconduct. Therefore, rather than taking an enforcement-heavy approach, the institution has opted for an exhortation-focused approach that prioritizes media literacy and appeals or exhortations. The institution's Academic Integrity Policy has been updated for clarity on what constitutes a breach at three different levels, ranging from negligence to flagrant pre-meditated breaches. This provides direction for all institutional members. A GenAI-specific strategy is developed by a working group under the ambit of the Provost's office. This working group brings together expertise from campus academic stakeholders and situates GenAI use within the normative discourse of academic citizenship, framing exhortations within this context. In July 2023, the university's Centre for Teaching and Learning hosted a study group to explore the subject of AI in the classroom and to develop guidelines for instructors (Moore, 2023). In addition to best practices for writing assignments in the GenAI-present ecosystem developed by one of its members (Doering, 2023), the study group issued guidelines for faculty (CBU CTL, 2023). This normative framework balances academic freedom, academic integrity, and privacy. The *grundnorm* or normative foundation is academic citizenship rather than just academic integrity. It must be noted that the use of the term "academic integrity" is itself a discursive shift from "academic misconduct;" a shift to "what we aspire to rather than what we would punish or want to avoid" (McNeill, 2022, p. 491). Academic citizenship situates academic integrity in the expectation that scholarship must contribute to the broader society. This conceptualization of academic citizenship engages what has come to be understood as academia's civic responsibilities to society (Macfarlane, 2007). It requires understanding that academics must harness learning to be community change agents. This discourse resonates with CBU as it is grounded in the Antigonish Movement (Johnstone, 2020). This Movement emerged in the early twentieth century as an initiative to mobilize grassroots community economic development in regions of Eastern Canada that were grappling with socio-economic decline. For CBU, therefore, GenAI misuse is not simply an issue of academic integrity; it is discursively framed as a practice that is harmful, to the individual and also to the collective.

CBU's *Charter of Academic Citizenship* (Calabrese et al., 2020) lists five values upon which the norm of academic citizenship is built: originality (individual intellectual efforts lead to new ideas and the growth of collective knowledge); integrity (as a core principle of academic life and acknowledging contributions to knowledge made by others); fairness (to peers and other academic community members); collegiality (mutual dignity and respect); and, responsibility (to maintain academic standards). These are informed by the definition of academic integrity by the

International Center for Academic Integrity (2021). But, just like any normative enactment, CBU's academic citizenship values must be read together with other institutional values, such as the list of attributes graduates are expected to demonstrate, including active and responsible citizenship (CBU, n.d.).

What does the misuse of GenAI mean for academic citizenship in the CBU context? Comments made by the study group participants in a Canadian Broadcasting Corporation interview (Moore, 2023) are insightful. For one, there is an issue of authenticity and originality that is in the realm of originality and integrity in the *Charter of Academic Citizenship;* there is a concern that AI offers opportunities for overreliance on technology to produce material that a student could pass off as their own. The other issue is one of critical thinking; CBU seeks to produce active change agents who are able to apply critical thinking skills to provide solutions to societal problems. The group commented that a slavish reliance on GenAI undermines critical thinking skills development.

Concurrent with the Centre for Teaching and Learning's study group, CBU's Writing Centre developed a student-focused Moodle module titled "AI & CBU Students." This module approaches GenAI from a critical thinking or intellectual curiosity perspective and allows students to learn about responsible GenAI use vis-à-vis institutional aspirational graduate attributes. It is noteworthy that the module is informed by media ecology. The introduction to the module states: "To be critical adopters of technology, users must question what new technologies offer us and what they take away" (CBU Writing Centre, 2023).

Organizational communication theory teaches that institutions are able to foster a preferred culture through, among other things, promoting values and using "heroes" as exemplars of institutional values (Miller, 2011, p. 33). CBU has adopted this through the creation of an annual award for students who "demonstrate commitment to the values identified by CBU's *Charter of Academic Citizenship*, especially Originality, Integrity and Collegiality" (CBU, 2022). The rationale is "[t]o recognize students who embody the highest standards of academic citizenship, and to continually encourage conversations about academic integrity" (CBU, 2022). The foregoing demonstrates CBU's prioritization of academic integrity within its cultural ecosystem.

Academic integrity is about taking responsibility and approaching one's work with truth and honesty. Media ecology considers the relationship that technology has on the environment within which it is used and how it affects society. To be an engaged citizen in an academic community from which one benefits, and to which one has responsibility, it is expected that one would approach the work they do ethically and with integrity, even while benefiting from academic freedom. We know from an ecological perspective that once a new technology is introduced, we do not have a new technology in the same society: we have a new society (Postman, 1992). As McLuhan's associate Father John Culkin said, "Life imitates art: We shape our tools and thereafter the tools shape us" (Culkin, 1967, p. 70). The implications this has for citizenship in a broad sense overlaps academic fields. GenAI tools can write and respond, but they are still not human; GenAI tools need human connection and relationship to thrive. Humans, too, need interpersonal connections to thrive.

In the context of academic citizenship, human relationship is not limited to the capacity to think critically, but also to feel. This is engaged by the self-reflection and comprehension requirements inherent in traditional literacy. GenAI platforms lack the ability to emote and feel for themselves, and therefore, are unable to critically reflect on experience due to the lack of empathetic reasoning (and said experience). They can only guess at best, and then mimic (Brookhouse, 2023) what the user may feel in any given situation. Not only is there a question of affective authenticity, but relying on a machine to feel for us bypasses the synapse of critical and reflective processes that deepen our understanding of experience and contribute to personal growth, and even, resilience (Walsh, 2025). For one of the authors, these concerns were demonstrated in a student paper about how to effectively express emotions. As an instructor, they were looking for demonstration of process-oriented critical thinking and critical self-reflection. Instead, they received an example of academic cheating and disregard of any review process. There was no question that the student engaged GenAI to produce this answer: "If I were a human, I may feel sad in this situation." There are two glaring issues with this submission: clearly, the student failed to engage the critical thinking capabilities or competencies needed to proofread and edit as necessary; but, it also proves that the GenAI tool cannot feel for itself, and therefore, cannot accurately engage in the critical self-reflection process that is necessary to answer the question. It confirms the limitations of GenAI that have been identified in other studies; despite its vaunted capabilities, the new technology cannot perform all tasks without the human control element, at least, yet (Tan et al., 2024). As a corollary, ethically, we have to consider the extent to which GenAI misuse normalizes (or even, glorifies) cheating (Walsh, 2025), with ramifications for society. For example, in a media ecosystem with a high potential for mis(dis)information, GenAI misuse could mean an abdication of academic fact-checking responsibility.

Conclusion

In the 1980s, Postman argued that the television as a medium has degraded political discourse, replacing print culture-driven reasoned public affairs with a culture of political spectacle and sensationalism (Postman, 1986). Today, is GenAI a dangerous invasive organism in the academic ecosystem? The proverbial jury is still out on the question (Cottom, 2025; Deng et al., 2024) and no doubt there will be multiple answers if the question is fed into a GenAI application. In this chapter, from a media ecological perspective, we have shared observations from a small Canadian university, identifying trends of GenAI use in the classroom, and CBU's approach. In the classroom, while GenAI misuse problematizes instructional gate-keeping power, GenAI is recognized as a new species within the media ecosystem that humans would have to consciously navigate through academic gatekeeping and media literacy. At the strategic-operational level, CBU has approached the GenAI question with a pragmatic, normative-exhortation approach that makes GenAI less of a frightening species than simply, a new one. It is an approach that recognizes students as partners in fostering academic integrity. After all, as Tan et al. (2024) note in their chapter on GenAI in the higher education context: "[A]ll stakeholders will need to work together to ensure that AI benefits outweigh the costs" (p. 146). The CBU approach is also consistent with the approach McNeill (2022) describes from their experience in the University of British Columbia, whereby the focus is placed on mainstreaming norms that undergird academic integrity.

However, from a media ecological perspective, environmental characteristics must be accounted for. In that sense, we must consider how academic community members can create an environment in which GenAI cannot wreak havoc. In a chapter contributed to *Academic Integrity in Canada*, Crossman (2022) provides a political economy rationale for the prevalence of academic integrity infractions: to the extent that higher education has become market economy-driven as government funding fails to keep the pace with inflation, "knowledge is seen as a commodity, transcripts and credentials as products, and students as consumers" (p. 217). They quote Westheimer's (2018) argument that universities have pivoted "from a mission of public knowledge and citizenship to a mission of job training and profitable patents" (p. 224). In light of *all* of the foregoing dynamics, the normative grounding of higher education in academic citizenship is sacrificed on the altar of transactional economic expediency. What does this mean for enrolment gate-keeping or quality control rigour? What are the expectations of the student-as-customer? What are the actual consequences for a student within the academic market economy if they are detected to have cheated? What are the ramifications of the rise in contract or sessional instructors in precarious employment and who are concerned about the impact of market measurement tools (like teaching evaluations and anonymous online ratings) on their job security if they are perceived as being rigorous by the academic customer audience? As Crossman (2022) argues: "This system sets the stage for a consumer-oriented approach to education, where the customer is always right, even when they are doing wrong" (p. 222). And if students feel that they are being objectified or commodified such that their genuine academic interests or well-being necessary to pursue rigorous academic work are not prioritized (Lundy, 2022), what responsibility will they feel towards academic citizenship? The subject of GenAI in the academy should, therefore, not be discussed in a vacuum; the characteristics and health of the ecosystem contribute to the effects of the new species within it.

References

Algra, K. (2021). Historical perspectives on plagiarism, borrowing and citation. In J. Soeharno & K. Algra (Eds.), *Plagiarism in academic research and education* (pp. 22–30). Association of Universities in the Netherlands.

Armstrong, L. (2022, December 30). N.S. university limits foreign student enrolment after recruitment too successful. *Global News*. Retrieved October 6, 2023, from https://globalnews.ca/news/9378983/cbu-international-student-limit-enrolment/

Ayers, T. (2023, October 13). Figures show Cape Breton University's growth far exceeds other Atlantic institutions. *CBC*. Retrieved October 14, 2023, from https://www.cbc.ca/news/canada/nova-scotia/cape-breton-university-explosive-growth-exceeds-other-atlantic-institutions-1.6995747

Baudrillard, J. (1988). *Selected writings*. Stanford University Press.

Brookhouse, O. (2023, May 23). Can Artificial Intelligence understand emotions? Retrieved October 14, 2023, from Telefónica Tech website: https://telefonicatech.com/en/blog/can-artificial-intelligence-understand-emotions

Buerger, G. E. (2021). Pens for hire. *Canadian Perspectives on Academic Integrity, 4*(2), 3–15. https://doi.org/10.11575/cpai.v4i2.74183

Calabrese, D., Russell, E., Hoover, J., & Byrne, T. (2020). CBU'S charter of academic citizenship. In *Academic integrity guide for students*. Cape Breton University. Retrieved from https://caul-cbua.pressbooks.pub/acadintegritycbu/chapter/a-charter-of-academic-citizenship-for-cape-breton-university/

Cape Breton University. (2022, September 1). Dr. Mary Keating Medal—Cape Breton University. Retrieved October 7, 2023, from Academic Integrity—CBU website: https://www.cbu.ca/future-students/academic-integrity/dr-mary-keating-medal-for-academic-integrity/, https://www.cbu.ca/future-students/academic-integrity/dr-mary-keating-medal-for-academic-integrity/

Cape Breton University. (n.d.). *CBU graduate attributes*. Cape Breton University.

CBU Centre for Teaching and Learning. (2023). *July 6_CTL guidelines for generative AI in teaching and learning.docx*. Cape Breton University. Retrieved October 6, 2023, from https://mycbu.sharepoint.com/:w:/s/CentreforTeachingandLearning460/EZqRXrShn51NudL7TsYT274BuB16l6zKzhwlqE2pzS8_mQ?e=fakcUk

CBU Writing Centre. (2023). Artificial intelligence & CBU students: Introduction. Retrieved October 6, 2023, from Cape Breton University website: https://cbulms22.cbu.ca/mod/book/view.php?id=195043

Choi, Y., & Hou, F. (2023). A comparison of postsecondary enrolment trends between domestic and international students by field of study. *Statistics Canada Economic and Social Reports, 3*(9), 1–25.

Cottom, T. M. (2025, March 29). The tech fantasy that powers A.I. is running on fumes. The New York Times. Retrieved May 12, 2025, from https://www.nytimes.com/2025/03/29/opinion/ai-tech-innovation.html

Crossman, K. (2022). Education as a financial transaction: Contract employment and contract cheating. In S. E. Eaton & J. Christensen Hughes (Eds.), *Academic integrity in Canada: An enduring and essential challenge* (pp. 217–230). Springer Nature. https://doi.org/10.1007/978-3-030-83255-1

Culkin, J. (1967, March 18). A schoolman's guide to Marshall McLuhan. *The Saturday Review*, 51–53, 70–72.

Deng, R., Jiang, M., Yu, X., Lu, Y., & Liu, S. (2024). Does ChatGPT enhance student learning? A systematic review and meta-analysis of experimental studies. *Computers & Education*, 105224. https://doi.org/10.1016/j.compedu.2024.105224

Doering, J. (2023, August 11). Writing-based assignments, artificial intelligence and academic integrity. Retrieved October 6, 2023, from CTL Blog website: https://medium.com/@cbuctl/writing-based-assignments-artificial-intelligence-and-academic-integrity-eac67111e2e5

Eaton, S. E., & Christensen Hughes, J. (Eds.). (2022a). *Academic integrity in Canada: An enduring and essential challenge*. Springer Nature. https://doi.org/10.1007/978-3-030-83255-1

Eaton, S. E., & Christensen Hughes, J. (2022b). Academic integrity in Canada: An enduring and essential challenge—Introduction. In S. E. Eaton & J. Christensen Hughes (Eds.), *Academic integrity in Canada: An enduring and essential challenge* (pp. xi–xvii). Springer Nature. https://doi.org/10.1007/978-3-030-83255-1

Eaton, S. E., & Christensen Hughes, J. (2022c). Academic integrity in Canada: Historical perspectives and current trends. In S. E. Eaton & J. Christensen Hughes (Eds.), *Academic integ-*

rity in Canada: An enduring and essential challenge (pp. 3–24). Springer Nature. https://doi.org/10.1007/978-3-030-83255-1

Eisenstein, E. L. (1983). *The printing revolution in early modern Europe*. Cambridge University Press.

García-Peñalvo, F., & Vázquez-Ingelmo, A. (2023). What do we mean by GenAI? A systematic mapping of the evolution, trends, and techniques involved in generative AI. *International Journal of Interactive Multimedia and Artificial Intelligence, 8*(4), 7–16. https://doi.org/10.9781/ijimai.2023.07.006

Hamilton, M., & Wolsky, K. (2022). The barriers to faculty reporting incidences of academic misconduct at community colleges. In S. E. Eaton & J. Christensen Hughes (Eds.), *Academic integrity in Canada: An enduring and essential challenge* (pp. 467–484). Springer Nature. https://doi.org/10.1007/978-3-030-83255-1

Innis, H. A. (1951). *The bias of communication*. University of Toronto Press.

International Center for Academic Integrity. (2021). *The fundamental values of academic integrity* (3rd ed.). International Center for Academic Integrity. Retrieved from https://academicintegrity.org/images/pdfs/20019_ICAI-Fundamental-Values_R12.pdf

Johnstone, H. (2020). Cape Breton University: Building on the CED legacy of Fr. Greg MacLeod. *International Journal of Cooperative Management Accounting and Management, 3*(1), 4–17. https://doi.org/10.36830/IJCAM.20201

Liang, W., Yuksekgonul, M., Mao, Y., Wu, E., & Zou, J. (2023). GPT detectors are biased against non-native English writers. *Patterns, 4*(7), 1–4. https://doi.org/10.1016/j.patter.2023.100779

Lundy, M. (2022, December 26). How Canada's foreign-student boom is creating a host of problems. *The Globe and Mail*. Retrieved October 15, 2023, from https://www.theglobeandmail.com/business/article-international-student-admissions-canada/

Lye, C. Y., & Lim, L. (2024). Generative artificial intelligence in tertiary education: Assessment redesign principles and considerations. *Education Sciences, 14*(6), 569. https://doi.org/10.3390/educsci14060569

Macfarlane, B. (2007). Defining and Rewarding Academic Citizenship: The implications for university promotions policy. *Journal of Higher Education Policy and Management, 29*(3), 261–273. https://doi.org/10.1080/13600800701457863

McLuhan, M. (1964). *Understanding media: The extensions of man*. McGraw-Hill.

McLuhan, M., & Leonard, G. B. (1967, February 21). The future of education: The class of 1989. *Look*, 23–25.

McNeill, L. (2022). Changing "hearts" and minds: Pedagogical and institutional practices to Foster academic integrity. In S. E. Eaton & J. Christensen Hughes (Eds.), *Academic integrity in Canada: An enduring and essential challenge* (pp. 487–503). Springer Nature. https://doi.org/10.1007/978-3-030-83255-1

Metz, C. (2023, May 1). 'The Godfather of A.I.' leaves Google and warns of danger ahead. *The New York Times*. Retrieved October 6, 2023, from https://www.nytimes.com/2023/05/01/technology/ai-google-chatbot-engineer-quits-hinton.html

Meyrowitz, J. (1985). *No sense of place the impact of electronic media on social behavior*. Oxford University Press.

Miller, K. (2011). *Organizational communication: Approaches and processes*. Cengage Learning. Retrieved March 6, 2023, from https://www.biblio.com/book/organizational-communication-approaches-processes-miller-katherine/d/1454194579

Milmo, D. (2023, June 23). Two US lawyers fined for submitting fake court citations from ChatGPT. *The Guardian*. Retrieved October 21, 2023, from https://www.theguardian.com/technology/2023/jun/23/two-us-lawyers-fined-submitting-fake-court-citations-chatgpt

Moore, M. (2023, July 4). CBU study group developing guidelines for instructors on AI software. *CBC*. Retrieved October 6, 2023, from https://www.cbc.ca/news/canada/nova-scotia/cbu-study-group-developing-guidelines-ai-software-1.6895734

Mumford, L. (1934). *Technics and civilization*. George Routledge.

Mumford, L. (1967). *The myth of the machine* (1st ed.). Harcourt, Brace & World.

Murray, M. (1989, April 21). Owner of essay-writing firm puzzled by sudden police raid. *Toronto Star*, A7.
Okerson, A. (1991). With feathers: Effects of copyright and ownership on scholarly publishing. *College & Research Libraries, 52*(5), 425–438. https://doi.org/10.5860/crl_52_05_425
Ong, W. J. (1982). *Orality and literacy: The technologizing of the word*. Routledge.
Postman, N. (1986). *Amusing ourselves to death: Public discourse in the age of show business*. Penguin Books.
Postman, N. (1992). *Technopoly: The surrender of culture to technology* (1st ed.). Knopf.
Richards, N. (2024). A modern Frankenstein: Professional military education reform in the age of ChatGPT. *Canadian Army Journal, 20*(2), 94–103.
Rolls, J. (2006). *Public speaking made easy* (2nd ed.). Nelson College Indigenous.
Standage, T. (2018). Social media retweets history. In P. Heyer & P. Urquhart (Eds.), *Communication in history* (7th ed., pp. 317–322). Routledge.
Strate, L. (2004). A media ecology review. *Communication Research Trends, 23*(4), 2–48.
Strate, L. (2008). Studying media as media: McLuhan and the media ecology approach. *MediaTropes, 1*, 127–142.
Tan, S., Rudolph, J., & Tan, S. (2024). Riding the generative AI tsunami: Addressing the teaching and learning crisis in higher education. In J. Rudolph, J. Crawford, C.-Y. Sam, & S. Tan (Eds.), *The Palgrave handbook of crisis leadership in higher education* (pp. 135–154). Springer Nature Switzerland. https://doi.org/10.1007/978-3-031-54509-2_34
Thompson, J. B. (1995). *The media and modernity: A social theory of the media*. Stanford University Press.
Valdez, J. (2023, October 19). Pras blames AI for his guilty verdict in conspiracy trial. Does he have the right culprit? *Los Angeles Times*. Retrieved October 30, 2023, from https://www.latimes.com/entertainment-arts/music/story/2023-10-19/rapper-pras-artificial-intelligence-ai-conspiracy-trial-appeal
Verrax, F., & Scheid, F. (2024, July 15). ChatGPT in education: Breakthrough or disruptive innovation? Retrieved January 7, 2025, from *The Conversation* website: http://theconversation.com/chatgpt-in-education-breakthrough-or-disruptive-innovation-225379
Walsh, J. D. (2025, May 7). Everyone Is Cheating Their Way Through College. New York Magazine. Retrieved May 11, 2025, from https://nymag.com/intelligencer/article/openai-chatgpt-ai-cheating-education-college-students-school.html
Wakefield, J. (2023, October 18). Is AI about to transform the legal profession? *BBC News*. Retrieved October 21, 2023, from https://www.bbc.com/news/business-67121212
Westheimer, J. (2018). Fatal distraction: Audit culture and accountability in the corporate university. In J. McNinch & M. Spooner (Eds.), *Dissident knowledge in higher education* (pp. 217–234). University of Regina Press.
Zhao, X., & Bishop, M. J. (2011). Understanding and supporting online communities of practice: Lessons learned from Wikipedia. *Educational Technology Research and Development, 59*(5), 711–735. https://doi.org/10.1007/s11423-011-9204-7

Part II
Pedagogy, GenAI, and Academic Integrity

Chapter 5
University Instructors' Reactions and Adjustments to Generative Artificial Intelligence

Martine Peters and Catherine E. Déri

Abstract Since the introduction of ChatGPT in 2022, the rapid evolution of generative artificial intelligence (GenAI) has forced university professors to question the risks of plagiarism in students' written productions. Although plagiarism is the appropriation of someone else's text or ideas which are then used without giving the original author the credit, the situation is quite different when a student uses GenAI to write part or all of an assignment. Machines are not currently recognized as authors, because AI tools cannot take authorial responsibility for generated text. In this paper, we explore how university instructors have reacted to artificial intelligence since the arrival of ChatGPT and if they have adapted their teaching and assessment tasks with plagiarism being redefined. We completed 29 interviews with instructors from Quebec universities regarding GenAI benefits and drawbacks for student learning and assessment. We present results that position instructors on a continuum of adjustment to new technologies and discuss changes to written assignment tasks to ensure the responsible use of GenAI by students.

Keywords Instructors · Faculty · GenAI · Student writing · Plagiarism

Introduction

In the previous chapters, GenAI was presented with its promises and dangers to the education system. In this chapter, the pedagogical implications of the introduction of GenAI in our education system will be discussed. In universities, instructors have

---

Instructors include lecturers and professors (sessional, teaching and/or tenure track professors, full professors, assistant professors, associate professors

---

M. Peters (✉) · C. E. Déri
Université du Québec en Outaouais, Gatineau, QC, Canada
e-mail: martine.peters@uqo.ca; catherine.deri@uqo.ca

A. E. King (ed.), *Artificial Intelligence, Pedagogy and Academic Integrity*, Advances in Artificial Intelligence in Education 1,
https://doi.org/10.1007/978-3-031-92534-4_5

a role to play in the promotion of academic integrity (Peters et al., 2019) and its values: honesty, trust, fairness, respect, responsibility, and courage (International Center for Academic Integrity, 2021). Instructors (for the purpose of this paper, instructors include lecturers and professors (sessional, teaching and/or tenure track professors, full professors, assistant professors, and associate professors)) need to ensure that the learning environment is equitable for all students, which includes having assessment tasks that are fair, impartial, rigorous, and valid. This will help students be responsible for their own learning and prepare them for their professional life. In their research, Guerrero-Dib et al. (2020) have shown that students who plagiarize during their studies tend to reproduce these fraudulent behaviors in their professional lives. Thus, it becomes imperative for instructors to promote academic integrity to guarantee the value of diplomas and degrees (Solmon, 2018) and, most importantly, to be able to certify that students have learned what they need to succeed in their studies and their profession.

Academic Integrity, Artificial Intelligence and Plagiarism in Universities

Academic integrity is and has been a concern in universities for well over a century (Aaron & Roche, 2013; Denisova-Schmidt, 2017; Gallant, 2008). Research on plagiarism has shown that cases are slowly rising in universities (Fass-Holmes, 2018; Jones, 2011) and have increased significantly in certain disciplines during and since the pandemic of COVID-19 (Hill et al., 2021; Kratovil, 2021; Thomas et al., 2022). Students have been caught plagiarizing or cheating, whether intentionally (Soppe, 2013) or unintentionally (Adam, 2016; Zwick et al., 2019). Moreover, the reasons behind these academic misconducts range from lack of time, laziness, pressure from parents or peers, to lack of knowledge and skills (Harji et al., 2017; Šprajc et al., 2017).

The arrival of GenAI in universities has resulted in an upheaval in the field of academic integrity, with many instructors, administrators, librarians, and even students claiming that GenAI can and will be used to plagiarize (Chan & Hu, 2023; Chan & Lee, 2023). Whereas plagiarism was previously defined as taking someone else's words or ideas to pass them off as one's own to gain a benefit (Jeffrey & Dias, 2019), using GenAI induced a change in the definition of plagiarism. According to certain authors, everything that is produced by GenAI that is not referenced is considered plagiarism (Koplin, 2023; Peters, 2023) because the words and ideas come from a database of works that were produced by individuals (Haque, 2023). In fact, Furze (2023) goes as far as saying that a student who copies and pastes from GenAI without referencing the original author is essentially plagiarizing a plagiarizer. Therefore, Peters (2023) redefined plagiarism after the arrival of GenAI, adding to the original definition that any text generated by GenAI is considered plagiarism if the use of GenAI is not permitted and not declared.

This new type of plagiarism has become a source of concern for all instructors in Higher Education (Yu, 2023) with some establishments going as far as banning GenAI tools (Barrett & Pack, 2023). Some instructors choose to warn students not to use GenAI to write their assignments, others choose to go back to pen and paper assessments (Yeo, 2023). However, many research findings have already been published showing the rapid adoption rate of GenAI writing tools in the student population (Hutson et al., 2024), and consequently an increase in student plagiarism using GenAI (Chan & Lee, 2023).

Need for Training about GenAI Tools

If students have been adopting GenAI tools, the same cannot be said about their instructors who are more wary about integrating GenAI in their tasks and showing students how to use it for their assignments (August et al., 2024). Many instructors have reported the lack of training offered (Barrett & Pack, 2023). This raises concern for two reasons. Firstly, instructors might not be aware of the potential of GenAI for learning, teaching, and plagiarizing because they are not very familiar with GenAI. A large variety of GenAI tools are now available for instructors to help improve learning outcomes and support them in developing better assessment practices that are less susceptible to plagiarism.

Secondly, those instructors who lack knowledge on the ways to use GenAI tools with integrity are not able to teach and educate students to properly use GenAI (August et al., 2024). New GenAI tools are being launched almost daily, many of which can help university students personalize their learning with AI tutors or content creators for example. A situation in which both instructors and students need training can lead to disastrous consequences where students use GenAI to plagiarize, intentionally or unintentionally, because they lack well-informed guidance from their instructors.

Need for Changes in Assessment Methods

The types of assignments asked of university students are often written essays (Graves et al., 2010). In their research, Kumar and Mindzak (2024, p. 6) show that ChatGPT "has proven to do a decent job of mimicking human writing." Therefore, instructors must change their assessments to evaluate students' progress and success so that they reflect what students, and not GenAIs, produce (Moorhouse et al., 2023). To modify their assessment practices, instructors must learn about GenAI and adjust their instructions and evaluation criteria. This will require an open mind from instructors to accept change by adapting to new ways of teaching and learning with up-to-date technologies that continue to evolve at a rapid pace.

Theoretical Framework: The Adjustment to New Technologies in Education

The theoretical framework that we used in this chapter was developed by Raby (2004) to describe the steps that instructors go through to adjust to new technologies. In educational environments, the four stages start with the "non-use" of a new technology then moves to the full "use" of the technology and are labeled: (1) awareness; (2) personal usage; (3) professional usage; and (4) pedagogical usage. Although some instructors will follow the steps in chronological order, there may be personal motivations or professional obligations that will result in steps being skipped completely or being reversed in their order of priority (Fig. 5.1).

In the **first stage** (awareness), the instructor is indirectly exposed to the new technology in a professional or personal setting (Raby, 2004). For example, the instructor is aware of colleagues, friends, or family members using and appreciating the new technology, but is not in direct contact with the software, application, or device. In the case of GenAI, it could be students sharing with their instructors how they experimented with different tools. This may raise enough awareness to spark the interest of the instructor to proceed to the **second stage** (personal use) and start using the new technology in their personal life. During this period, the instructor is motivated to explore technical functionalities without developing a full understanding of operating systems (Raby, 2004). This leads to the appropriation of the new technology, in a sense that it is regularly used to meet personal needs: searching for information, communicating with friends and family members, or generating a variety of digital products (i.e., documents, audios, videos, etc.).

Then, due to professional obligations or on a needs basis, the instructor will enter the **third stage** (professional use), during which they will use the new technology for work (Raby, 2004). They will further their familiarization of functionalities to assume responsibilities related to a specific job. In the case of university professors, they may use the new technology to facilitate research or administrative tasks.

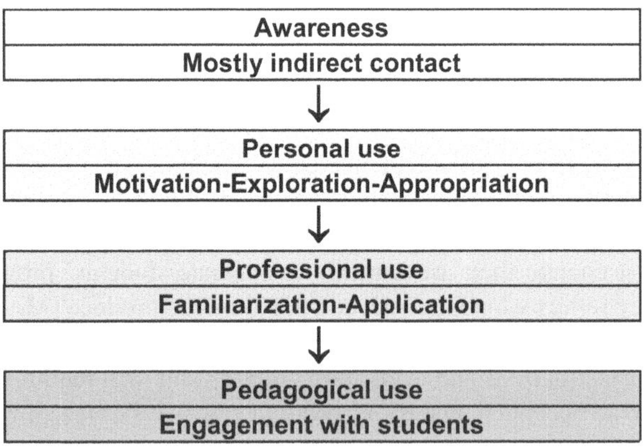

Fig. 5.1 Model of adjustment to technology (Raby, 2004)

Finally, the **fourth stage** (pedagogical use) is again triggered by intrinsic or extrinsic motivators (curiosity, job requirements, or other obligations). The level of comfort that instructors develop with technology (i.e., previous digital experiences in educational settings), can also influence individual situations (Raby, 2004). During this stage, instructors will use new technology to complete tasks that are directly linked to teaching and learning activities involving students. Therefore, instructors are not only using the new technology to plan such activities, but they are also designing learning experiences during which students are active participants in developing practical skills for their professional lives after their studies.

According to Raby (2004), instructors who manage to make the best use of new technologies are those who have dabbled in each of the stages of the theoretical model; thus, using said new technology to meet both professional and personal requirements. Eventually, those instructors who spend enough time in the fourth stage will make an exemplary use of the new technology through its regular and frequent application (Lefebvre et Fournier, 2014; Morais, 2001; Parks et al., 1994). In an educational setting, the term "exemplary" means that the instructors develop a level of proficiency serving as models for other instructors and students to emulate (Pierson, 1999). Although Raby's theoretical framework was developed to analyze the reactions of primary school teachers, we believe it can be applied to other levels of education, including how university instructors reacted to the advent of GenAI in the year following the launch of ChatGPT.

The aforementioned proficiency can only be attained by reflecting on how to maximize the learning outcome for students, such as by designing meaningful assessments. In that respect, Lodge et al. (2023) developed a taxonomy of six options to redesign assessment for GenAI: Ignore; Ban; Invigilate; Embrace; Design around; and Rethink. The first option (Ignore) is to dismiss the new technology, seeing it as just another digital trend that will not have a lasting impact on education. The second option (Ban) is to prevent the use of the new technology, which is becoming increasingly challenging since GenAI functionalities are integrated into commonly used software and applications. The third option (Invigilate) is to design assessment in a way that avoids the use of GenAI, such as reverting back to more traditional evaluation strategies. The fourth option (Embrace) is to integrate the new technology in assessments by "allowing or requiring students to use GenAI in specific tasks" (Lodge et al., 2023, para 8). The fifth option (Design around) is to think about assessment that could exploit the limitations of the new technology, which, according to Lodge et al. (2023), may not be a longstanding approach due to ongoing improvements to GenAI. Finally, the sixth option (Rethink) is to challenge how and why students are being assessed, ensuring that evaluations are meaningful and maximize learning for future aspirations.

It is through the theoretical lenses of Raby (2004) and Lodge et al. (2023) that we examined the adjustment of university instructors to GenAI in institutions located in the French-speaking province of Quebec in Canada. We were primarily interested in better understanding where they were situated in Raby's model of adjustment to technology, one year after the launch of ChatGPT. We also wanted to identify the changes implemented to their assessment strategies to maximize the learning experience of students and/or to potentially limit plagiarism in students' assignments.

Methodology: A Mixed Design International Research

Our research project consists of an international partnership focused on university plagiarism prevention involving 63 scholars from 32 institutions across North America and Europe. The main objective is to yield new knowledge on competencies leveraged, learned, and taught during undergraduate programs to prevent plagiarism in written assignments. We developed data collection instruments based on a theoretical framework by Peters (2015), categorizing digital strategies into three types of skills: informational, writing, and referencing. During our seven-year (2021–2028) mixed methods study, there are three phases of data collection: (1) questionnaires, (2) individual interviews and focus groups; and (3) recorded writing tasks. In this section, we will solely focus on individual interviews conducted with university instructors working at universities in Quebec.

During the period from November 23rd, 2023, to May 3rd, 2024, we conducted individual interviews with a total of 29 professors coming from six different universities. These participants identified to the following gender categories: female ($n = 16$), male ($n = 12$), and non-binary ($n = 1$). Their average age was 47 years old (ranging from 29 to 71 years of age). They also specialized in 15 distinct disciplines, with an average of 13 years of teaching experience (ranging from 3 months to 35 years). There is a difference in proportion of status between Lecturers (12%) and Professors (88%), although among the latter category only five participants identified themselves as Full Professors, whereas others indicated that they were employed in Assistant Professor, Associate Professor or simply Professor positions. Only 32% of the participants recalled receiving training in academic integrity from their affiliated institutions. In Table 5.1 below, sociodemographic details are provided on the convenience sample.

The semi-structured interviews were 38 minutes and 13 seconds on average in duration. The interview guide included 49 questions divided into five sections: (1) sociodemographic data; (2) informational skills; (3) writing skills; (4) referencing skills; and (5) general knowledge about plagiarism. The majority of interviews were done in person, but to facilitate the participation of some participants, a restricted

Table 5.1

Characteristics	Number of participants or description
Countries of origin	Canada: 24/Others: 5 (Algeria, Cameroun, Chili, Iran, Mexico)
Universities of affiliation ($n = 6$)	Université de Montréal, Université du Québec à Chicoutimi, Université du Québec à Montréal, Université du Québec en Outaouais, Université du Québec en Abitibi-Témiscamingue, Université Laval
Disciplines ($n = 15$)	Linguistics (English & French), History & Geography, Musicology, Industrial Design, Education, Psychology, Accounting & Administration, Nursing, Communications, Law, Political Sciences, Management, Sociology, Engineering, Computer Sciences
Academic integrity training	Yes: 9/No: 15/Uncertain: 5

number of instructors were met online using videoconferencing services (Zoom). The audio recordings were transcribed to facilitate a content analysis with the NVivo software. To do this, we focused on answers to questions from the last section of the interview guide pertaining to the instructors' knowledge, usage, and integration of AI tools. Since the interviews took place 12 to 18 months after the launch of ChatGPT by OpenAI, the interviewers prompted participants to comment on this specific GenAI tool, while also encouraging contributions about other types of AI. Finally, we analyzed responses from participants related to their adjustments made to instructions and evaluations of written assignments, since the onset of the recent GenAI upsurge.

Results

The results show that, regarding GenAI adoption, some instructors follow Raby's (2004) proposed stages. However, for various reasons, some do not and choose to reject or avoid GenAI. An adaptation of Raby's (2024) model is presented below including two additional stages (rejection and avoidance) that emerged from our data analysis (Fig. 5.2).

As part of our analysis, we categorized the instructors by stages of adjustment according to their most frequent mentions of strategies associated with specific stages of the model. However, instructors may be associated with a particular stage, for example the "professional use" stage, because most of their comments related to that stage, but they may have also expressed some comments linked to other stages in the model. Of the 29 instructors who were interviewed, the distribution per stage was the following: Rejection (n = 5); Avoidance (n = 3); Awareness (n = 8);

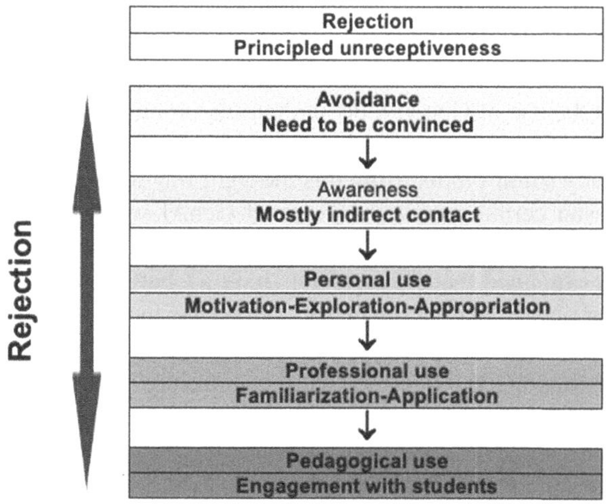

Fig. 5.2 Model of adjustment to GenAI. (Adapted from Raby (2004))

Professional use (n = 5); and Pedagogical use (n = 8). None of the participants reported being at the personal use stage, in part because we did not ask questions about the instructors' utilization of GenAI at home. The results presented below will highlight the reasons which prompted instructors to situate themselves at one stage or another of the model of adjustment to GenAI.

No Adjustment: Principled Rejection of GenAI

A total of five participants in our study rejected the use of GenAI. This rejection meant that, for them, this new technology was **not an improvement** to their personal and professional lives. One participant reflected that "one of the drawbacks (…) was the way in which AI was introduced, by the newspapers, as a threat to our lives as teachers, as learners" (Participant 017). The participant further explained that GenAI can seem to be more capable than humans, which can be quite worrisome for people.

One of the main reasons mentioned by the instructors who choose to reject GenAI is the possibility that it will **impede student learning**. Participant 004 pointed out that "it's important to start from scratch and to understand how to structure your speech, your thinking, how to communicate, and by having these tools, I feel like [students] are skipping a step that is so important in so many facets of their lives." Another instructor remarked that "GenAI is a bit of a shortcut to analysis and that's going to put our students at a disadvantage, because they're going to be dependent on GenAI to stand out from the crowd in regard to their analyzing abilities" (Participant 018). As for Participant 015, he made it clear to students that their main competitor will be GenAI and that his role was to help them develop the skills needed to do a better job than GenAI.

An obvious reason why some instructors choose to reject GenAI is the way it can be **used to plagiarize** in Higher Education. Many participants in our study were worried that all students are now plagiarizing using GenAI. Some have chosen to completely ban the use of GenAI tools in their classrooms while others have gone farther, prohibiting the use of GenAI in their study programs "to send a message to our students that we don't think using it is the right way to go" (Participant 018).

The last reason certain instructors rejected GenAI was **based on principles**: environmental concerns, refusal to encourage megacorporations, and data security. Participant 016 explained that she does not "have a ChatGPT account because this is a type of surveillance, that they watch all that we do and I make sure to avoid this type of problem." Another instructor (Participant 014) was quite adamant that all technologies, including GenAI, are "killing relationships, families, democracy, and schools." He even told his colleagues that "while you send your money to Silicon Valley, the culture in Quebec is the one that suffers" (Participant 014).

Even though some participants outright rejected GenAI with a complete refusal to engage with the new technology from the start, others engaged in other stages of adjustment (Raby, 2004) before choosing to partially or completely reject GenAI.

For instance, Participant 002 who reached the stage of pedagogical use still proclaimed that "if we become too dependent, if we rely too much on GenAI, it is a shortcoming." Participant 009, while at the stage of professional use, stated that "in Accounting, lots of tasks are repetitive and this gives a chance to students to build their expertise, but if ChatGPT is the one that does these tasks, students are going to leave school without going through these steps and without acquiring the skills they need. That scares me." These two examples clearly show how even if some participants have initially adopted GenAI, they choose to reject it in certain circumstances and for specific tasks, to maximize their students' learning experience.

One Step Prior to Adjustment to Technology: Avoidance of GenAI

Although Raby's (2004) model starts with the awareness stage, our findings show that there is a previous stage of avoidance to which three participants identify. This stage is different from rejection because the participants who choose to avoid GenAI at this time, remain open to adopting the new technology at some point in the future. "So for the moment, my view isn't clear-cut enough, I don't want [the students] to use it. Right now, it's a problem! I'm open to changing my mind eventually" (Participant 029). Similar participants explained that they avoid using GenAI based on four reasons that are also shared by other instructors, at various stages, depending on the circumstances.

The first reason is comparable to the initial one at the rejection stage: the negative impact of GenAI on **the development of students' skills**. A participant stated that "I think it's something that's uncontrolled and extremely frightening for the development of intelligence in young and old alike. That's why I've gone back to handwriting" (Participant 011). The second reason is that some participants **feel overwhelmed** by the tools themselves (complexity, quantity, evolution), but also by the future of Education (change, uncertainty). One participant said, "I am positive that there will be an impact, but I am still unsure of what it will be, what form it will take because of the rapidity of the change, it's really impossible to predict" (Participant 001).

Yet another reason mentioned is the **amount of work required** to implement GenAI. Participant 003 pointed out that "I had a big existential crisis when ChatGPT came out, you know. Now, I mean, I'm out of the big crisis and I think there are positive things to be done with it, but it's going to take a lot of collective work." Finally, some instructors avoid the new technology because they **need training and direction** (policies) from their institutions. Participant 024 stated that "there are questions I'd like to get answers to, just as I'd like a clearer policy on what we can do at my university in this area. I know that everyone's a bit behind the times when it comes to technology. Technologies are advancing much faster than what we're able to produce as a framework, but I have to admit that, as a new teacher, I feel a bit lost and I lack resources."

First Step of Technology Adjustment: Awareness toward GenAI

Many of the participants (n = 8) are at this stage of awareness toward the new technology, not surprisingly because the arrival of GenAI in Higher Education is still recent. Some instructors have already **tried some of the tools**, because they were curious and wanted to understand how they worked. For example, Participant 001 mentioned: "I tested it once to try to understand this thing. It's easy to learn how to use the tool." Several instructors want to try GenAI to see what it can do and if students can use it to write the assignments they are requesting in their classes. "I just tested [the tools] to see if they had been used by my students. It was the first time I used them because I don't really master this technology, so I'm not too sure what to do with it" (Participant 024).

Some of the instructors mentioned that it is their responsibility to explore GenAI, by **seeking training** to be able to understand how these tools can help them in their work. "I attended a presentation where a speaker explained how he was using it to send automated emails that were personalized to each addressee according to the response. I thought it was a good idea because it could prevent email overload and burn-out" (Participant 012). Others explained that they needed to understand GenAI to demonstrate that they knew what it was all about. "I attended a conference we had last spring. The speaker just showed us a few tools. I know I have to use them anyway to better understand. I want to put them to good use. I don't want to be a bit of a dimwit" (Participant 013).

At this stage, most participants were willing to try GenAI tools to **understand but also embrace them**. "I participated in some kind of training on ChatGPT and other tools to ensure students didn't plagiarize with ChatGPT. I have the impression that you have to learn to live with it" (Participant 019). The same sentiment is shared by Participant 022: "I looked at the PUPP [Partnership on University Plagiarism Prevention] website, and I read certain articles to see how things worked at colleges or universities. I quickly came to the conclusion that it's inevitable. It's like Google or something, so we have to integrate it". These participants will promote GenAI use with their students but also integrate it in their own tasks.

Second Step of Technology Adjustment: Personal Use of GenAI

None of our participants were categorized at this stage of the theoretical framework. This can be explained in two ways. Firstly, none of the questions posed during the interview specifically addressed the use of GenAI in personal lives. The second explanation is that, at this point, GenAI has taken on the form of a large menace for

Academia, due in part to the alarmist way the media has portrayed it for Higher Education. This is probably why instructors have a lot more to say about their use of GenAI in professional settings than in their personal lives, as can be seen in the next section.

Third Step of Technology Adjustment: Professional Use of GenAI

Among our participants, five were at the stage of professional use of GenAI but many others also mentioned using GenAI for various tasks in professional settings. Many participants report using GenAI to **search for information**, such as Participant 002 who explained that "ChatGPT is a search engine. It can be a pretty useful resource if you know how to use it and have the right intentions, to make a list of references for example." Participant 009 described ChatGPT as a librarian that is always available and most of the time reliable. As for Participant 007, he indicated that ChatGPT is:

> An open tool, like Wikipedia, like our databases, the specialized stuff. I mean, for a lot of questions… Very, very simple everyday examples: I'm reading a text, I can go and open an online dictionary, but if I feel I need a context, what does ChatGPT know about it? I'm not saying that all of a sudden, ChatGPT becomes my reference, but it's because I find it interesting. In fact, it adds to it. That is, it doesn't really act as a replacement, but it adds to the rest. Sometimes, it also allows me to control its level of quality. Because if I want to be able to - and at first it was conscious, now it's integrated into my daily tasks - if I want to be able to talk about it to my students, I have to use it, in high doses.

Another instructor (Participant 006) used GenAI for **programming** and as a tutor in statistics. Participant 017 **created videos and images** with GenAI tools while Participant 012 found it useful to **write emails**. However, it is surprising that none of the participants mentioned using GenAI to write academic texts. Is it possible that the instructors perceived a stigma attached to using GenAI to write and so they avoided talking about it? Some did explain how they used it as a tool to help with their writing, for example as a **translation** tool, but not for writing. Participant 017 commented that GenAI is useful for **revising and restructuring** a text, and Participant 026 used it to generate creative titles. Participant 007 described how she used it to **brainstorm** for her writing: "We're in the process of producing a dictionary in our field and for many of the entries, we test entries, we test sub-sections of texts, we ask ChatGPT 'What do you think of this, define this entry, that entry'. So, we use it to compare vocabulary to see what's going on and sometimes it gives us some very interesting ideas".

Fourth Step of Technology Adjustment: Pedagogical Use for GenAI

A total of eight participants were categorized at the stage of pedagogical use of GenAI and they mentioned four different teaching strategies. The first type of pedagogical use was the **planning of courses and activities** for students. Participant 013 explained that she used GenAI to formulate an AI usage directive for her classes that she inserted in the syllabus. Students were then aware of what they were allowed and not allowed to do with GenAI when writing assignments. Another instructor mentioned that he used ChatGPT to create simulations: "I provide ChatGPT with certain contexts and ideas and request a table of content. I get ideas that I might not have thought of, so I don't have to look elsewhere" (Participant 009).

Pedagogically, GenAI was also used for the prevention of plagiarism in two ways: to **modify existing assignments** so that they would be more difficult to plagiarize and to prepare activities to **show the potential but also the shortfalls** of GenAI to students. Participant 003 described what a lot of the other participants expressed when she stated:

> I realized right away that I was going to have to change things because some of my teaching practices left the door wide open for the use of ChatGPT, and on a metaphysical level, I refuse to waste my limited time spent on this earth marking things done by a machine in six seconds. That's a no. In fact, I've systematically modified my exams and assignments so that it's impossible to do them all with ChatGPT.

Participant 003 also mentioned that she added a meta-reflection to her assignments and has eliminated take-home exams, while Participants 009 and 022 now request bibliographies from students to see where they have taken their information. Oral presentations are a lot more present in one of the participants' classes: "After students hand in a paper, they have a 15-minute presentation to do during which I question them to see if they really understood the study material. In the course outline, I specify that if the oral presentation is not adequate, I can revise their grade and give them a zero if I find that they have plagiarized" (Participant 013). Another instructor explained that teaching and assignments now have to be more contextualized and personalized, with more follow-ups to give students better guidance (Participant 021).

Another way to pedagogically integrate GenAI to prevent plagiarism is to show students how the new technology can be an ally but also their enemy, depending on how they use it. The instructors will facilitate effective GenAI use by their students. Participant 017 compared it to any tool:

> They're really in a learning mode, so I said to myself, maybe it's time to develop some training to help them use it properly. I really see it like anything else: a knife in the hand of a butcher, makes good packs of meat, but a knife in the hand of a thief, does something else. So, I see it as a knife and I want it to be in the hands of a butcher. That's how I see it.

While delivering courses to future teachers, Participant 020 promoted awareness by "showing them how ChatGPT worked, explaining how to use it adequately rather than forbidding its use. At the same time as I'm making them understand what not to do in my class, I'm raising their awareness, because they're going to be in my situation with their own students later on, and they'll have to find ways to deal with it." Three participants (007, 023, 025) generated a text with GenAI and analyzed it with their students to show how the generated text was not appropriate or adequate. "I actually showed them ChatGPT's shortcomings to convince them that it's a tool and not a substitute for human intelligence, that it doesn't actually replace the effort of understanding and writing" (Participant 025).

The fourth and last pedagogical use of GenAI was to **detect plagiarism** in students' assignments. Although some instructors complained about the fact that the technology cannot, at this time, reliably detect GenAI in written assignments, a few participants still used various tools to check for plagiarism. Participant 024 decided to verify an assignment "because I had suspicions about a text that was well written, in fact, relatively impeccable in terms of argument, but had certain repetitions that were a bit off and had no references. I had my doubts. I used [GenAI] to check". This pedagogical use of the new technology was not often mentioned by the participants.

Discussion

The results show that our participants are not all at the same stages of adjustment to GenAI but that the majority of them (72%) have adopted it at the awareness, professional, and pedagogical stages. The rest of the instructors are at the rejection or avoidance stages (28%). Using these findings, we illustrate the links between the stages of adjustment that we adapted from Raby's model (2004) and the taxonomy of options to redesign assessment for GenAI by Lodge et al. (2023). The personal and professional stages have been eliminated from this diagram because they are not associated with pedagogical tasks (Fig. 5.3).

Pedagogical Choices of Rejectors

The instructors who are at the rejection stage can choose two of Lodge et al.'s (2003) assessment strategies. They can simply **ignore** GenAI and adopt the *status quo* for their pedagogical practices. However, Lodge et al. (2023) specify, "this approach seems unlikely to be viable in the long term" (para 6). Because Rejectors refuse to use GenAI, their second strategy will be to **ban** it from their classes and request that their institution ban it completely. Again, this is not a viable strategy because students will find ways to outmaneuver the restrictions (Lodge et al., 2023). The problem with the Rejectors' stance is that they will likely be going in a different direction than the majority of their colleagues and may offend many students by their unwillingness to compromise. This will lead to a climate of dissatisfaction and distrust where instructors and students are not working together to maximize learning.

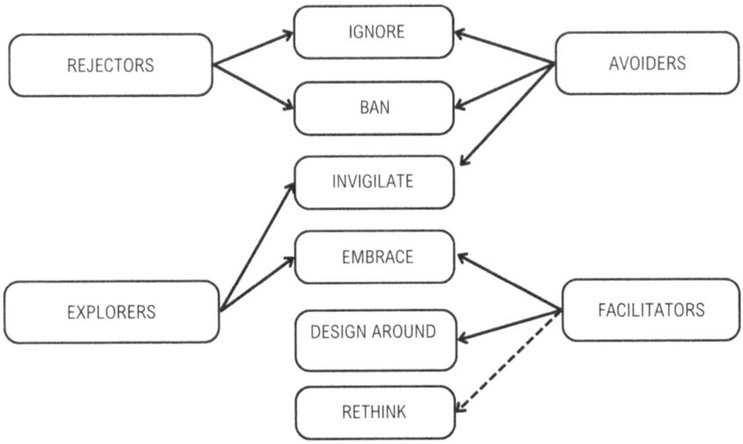

Fig. 5.3 Proposed model of adjustment stages of redesigning GenAI assessments. (Adapted from Raby (2004) and Lodge et al. (2023))

Pedagogical Choices of Avoiders

Like Rejectors, Avoiders, who are basically uncertain about GenAI, will most likely **ignore** it until they receive training or can refer to an institutional policy to determine whether to adopt or reject the new technology. However, some Avoiders remain concerned about GenAI and if it will be beneficial or detrimental to teaching and learning. With students' best interest in mind, Avoiders may also choose two other pedagogical strategies when it comes to their assessment. Those who believe that GenAI is detrimental to learning might choose to **ban** it in their classroom. Others might **invigilate** by circumventing the use of GenAI for their assessments to ensure students focus on course content and do not use GenAI to plagiarize. Some of our participants said that they were going back to more "traditional" practices, such as in-class, pen and paper assignments. While this approach shows an openness to adapting to change rather than a willingness to adjust to GenAI, it most likely will not be a winning strategy for students who seldom write without a computer nowadays and prefer to have technological tools at their disposal.

Pedagogical Choices for Instructors at the Awareness Stage: Explorers

When instructors are at the awareness stage, they are curious and basically explore the possibilities of GenAI. Some will choose, like the Avoiders, to modify their assignments so that students cannot use GenAI for their completion: the **invigilate** option proposed by Lodge et al. (2023). Other Explorers will **embrace** GenAI because they feel they have no choice but to accept its existence. They also wish to test the new technology to see its potential for teaching and learning. They are open to exploring GenAI on their own and with their students but remain cautious because of the issues mentioned by Lodge et al. (2023): "ethics, fairness, and equity, particularly in terms of privacy, access to advanced AI technologies, and the varying abilities of students to use them effectively" (para 9).

Pedagogical Choices for Instructors at the Pedagogical Use Stage: Facilitators

Instructors at the stage of pedagogical use are actively engaged in modifying their assessments to encourage students using GenAI in the learning process. These instructors have **embraced** GenAI, but they have also gone one step farther. They chose to facilitate the use of GenAI by students by teaching them the pros and cons of these tools. These instructors-facilitators are not exactly modifying their assignments by **designing around** to exploit the limitations of GenAI (Lodge et al., 2023). Instead, our instructors-facilitators are informing students of the limitations and the potential dangers of GenAI to academic integrity if they are not using the new technology properly when writing assignments. Facilitators are providing training to students on how to properly complete assignments, with or without GenAI, to maximize their learning and prevent plagiarism.

Finally, a few of our instructors-facilitators indicated that they might be completely **rethinking** their assignments. They were questioning how to better guide or offer support to students. This meant figuring out how to fully integrate GenAI in the teaching and learning process. However, very few of our participants had reached this point, partly because of the time and effort needed to engage in this kind of reflection (Lodge et al., 2023).

Challenges and Opportunities with the Adjustment to GenAI

At the time that we conducted the interviews with instructors, they were still facing three main challenges to best adjust to GenAI. First, they were generally lacking training opportunities, scrambling to find webinars, workshops, and conferences on the topic of teaching and learning in the new GenAI era. Those delivering these training sessions had developed recent and uneven expertise, focussing on general issues and providing advice on strategies that were not yet tested and proven. This is why several participants expressed a requirement for their institution to deliver tailored training.

Furthermore, the scientific literature was grossly lacking, forcing instructors to refer to less rigorous publications, such as blogs, websites, and media outlets to find information. Even though the information that was published in open sources was useful and thought provoking, it was still not meticulously developed through usual peer reviewed articles. Interestingly, one of our participants decided to make GenAI the focus of his research on technopedagogy. Other specialists in the fields of Education and Academic Integrity also shifted their research interests to investigate this new technology from teaching and learning perspectives, which will soon produce a more robust body of literature on the topic.

Lastly, the lack of institutional policies represented a challenge for several instructors who were hesitant to take initiatives that would potentially have to be overturned, depending on the directions that local administrators would eventually take. However, other instructors appreciated the flexibility to make their own decisions about the extent of the use of GenAI in their respective classes. Although academic freedom is important in Higher Education, a program approach should be adopted to ensure that the integration of GenAI is progressive and tailored to the needs of instructors, as well as students who are registered in the same study programs, so that they face consistencies in their learning experiences. Ultimately, the integration of GenAI should be aligned with the mission of Higher Education institutions which is to develop skills and knowledge in our student. This also means that developments in GenAI technology should be followed closely to best prepare students for typical work environments in their discipline.

Conclusion

The aim of this chapter was to examine how instructors are reacting and adjusting to the introduction of GenAI in Higher Education. The findings show that many instructors are reluctant to adopt GenAI tools for reasons such as lack of time, training, and resources, as well as an absence of institutional policies. However, there are instructors who are conscious of the necessity for change in order to properly integrate GenAI with integrity in our institutions. What is needed now, and rapidly, is training for the instructors, not to convince them of the importance of adopting GenAI tools but rather to demonstrate how pedagogical choices should be made to maximize students' learning. Future studies should focus on the evolution of instructors' adjustments to GenAI, with time and ongoing changes to the new technology. Moreover, pedagogical strategies, including assessment practices, should be examined to determine which teaching approaches will enhance students' learning with or without the use of GenAI.

References

Aaron, L. S., & Roche, C. M. (2013). Stemming the tide of academic dishonesty in higher education: It takes a village. *Journal of Educational Technology Systems, 42*(2), 161–196. https://doi.org/10.2190/ET.42.2.h

Adam, L. (2016). Student perspectives on plagiarism. In T. Bretag (Ed.), *Handbook of academic integrity* (pp. 519–535). Springer Science+Business Media.

August, E. T., Anderson, O. S., & Laubepin, F. A. (2024). Brave new words: A framework and process for developing technology-use guidelines for student writing. *Pedagogy in Health Promotion, 10*, 187–196. https://doi.org/10.1177/23733799241235119

Barrett, A. & Pack, A. (2023). Not quite eye to A.I.: Student and teacher perspectives on the use of generative artificial intelligence in the writing process. *International Journal of Educational Technology in Higher Education, 20*(1). https://doi.org/10.1186/s41239-023-00427-0

Becker, H. J. (1994). How exemplary computer-using teachers differ from other teachers: Implications for realizing the potential of computers in schools. *Journal of Research on Computing in Education, 26*(3), 291–321.

Chan, C. K. Y., & Hu, W. (2023). Students' voices on generative AI: Perceptions, benefits, and challenges in higher education. *International Journal of Educational Technologiy in Higher Education, 20*(43). https://doi.org/10.1186/s41239-023-00411-8

Chan, C. K. Y., & Lee, K. K. W. (2023). The AI generation gap: Are Gen Z students more interested in adopting generative AI such as ChatGPT in teaching and learning than their Gen X and Millennial Generation teachers? Smart Learning Environment, 10(60). https://doi.org/10.1186/s40561-023-00269-3

Denisova-Schmidt, E. (2017). The challenges of academic integrity in higher education: Current trends and prospects. *CIHE Perspectives, 5*, 1–27.

Fass-Holmes, B. (2018). International students reported for academic integrity violations: Demographics, retention, and graduation. *Journal of International Students, 7*(3), 644–669. https://doi.org/10.32674/jis.v7i3.292

Furze, L. (2023). Generative AI, plagiarism, and "cheating". https://leonfurze.com/2023/09/20/generative-ai-plagiarism-and-cheating/comment-page-1/

Gallant, T. B. (2008). *Academic integrity in the twenty-first century. A teaching and learning imperative*. Wiley Periodicals.

Graves, S. M. (2008). Student cheating habits: A predictor of workplace deviance. *Journal of Diversity Management (JDM), 3*(1), 15–22. https://doi.org/10.19030/jdm.v3i1.4977

Graves, R., Hyland, T., & Samuels, B. M. (2010). Undergraduate writing assignments: An analysis of syllabi at one Canadian college. *Written Communication, 27*(3), 293–317. http://wcx.sagepub.com/content/27/3/293.abstractN2-

Guerrero-Dib, J. G., Portales, L., & Heredia-Escorza, Y. (2020). Impact of academic integrity on workplace ethical behaviour. *International Journal for Educational Integrity, 16*(1). https://doi.org/10.1007/s40979-020-0051-3

Haque, M. A. (2023). A brief analysis of "ChatGPT" – A revolutionary tool designed by OpenAI. *EAI Endorsed Transactions on AI and Robotics, 1*(1), e15. https://doi.org/10.4108/airo.v1i1.2983

Harji, M. B., Ismail, Z., Chetty, T. N., & Letchumanan, K. (2017). Technical and non-technical programme students' attitudes and reasons for plagiarism. *English Language Teaching, 10*(11), 141–155. http://doi.org/10.5539/elt.v10n11p141

Hill, G., Mason, J., & Dunn, A. (2021). Contract cheating: An increasing challenge for global academic community arising from COVID-19. *Research and Practice in Technology Enhanced Learning, 16*(1), 24. https://doi.org/10.1186/s41039-021-00166-8

Hutson, J., Plate, D., & Berry, K. (2024). Embracing AI in English composition: Insights and innovations in hybrid pedagogical practices. *International Journal of Changes in Education, 1*(1), 19–31. https://doi.org/10.47852/bonviewIJCE42022290

International Center for Academic Integrity. (2021). *The fundamental values of academic integrity* (3rd ed.). https://academicintegrity.org/resources/fundamental-values
Jeffrey, D., & Dias, W. (2019). Perceptions du plagiat par de futurs enseignants québécois: dimensions éthiques et pédagogiques. *Revue canadienne de l'éducation, 43*(2), 767–790.
Jones, D. L. R. (2011). Academic dishonesty: Are more students cheating? *Business Communication Quarterly, 74*(2), 141–150. https://doi.org/10.1177/1080569911404059
Koplin, J. (2023). Plagiarism, academic ethics, and the utilization of generative AI in academic writing. *International Journal of Applied Philosophy, 37*(2), 17–40. https://doi.org/10.5840/ijap2023372202
Kratovil, A. (2021). Plagiarism in the graduate nursing program: Occupation stress or lack of knowledge? *Nursing Science Quarterly, 34*(4), 374–377. https://doi.org/10.1177/08943184211031599
Kumar, R., & Mindzak, M. (2024). Who wrote this? Detecting artificial intelligence–generated text from human-written text. *Canadian Perspectives on Academic Integrity, 7*(1). https://doi.org/10.55016/ojs/cpai.v7i1.77675
Lefebvre, S., et Fournier, H. (2014). Utilisations personnelles, professionnelles et pédagogiques des TIC par de futurs enseignants et des enseignants. *Revue internationale des technologies en pédagogie universitaire, 11*(2), 38–51. https://doi.org/10.7202/1035634ar
Lodge, J. M., Howard, S., & Broadbent, J. (2023). *Assessment redesign for generative AI: A taxonomy of options and their viability.* LinkedIn.
Moorhouse, B. L., Yeo, M. A., & Wan, Y. (2023). Generative AI tools and assessment: Guidelines of the world's top-ranking universities. *Computers and Education Open, 5*, 100151. https://doi.org/10.1016/j.caeo.2023.100151
Morais, M. A. (2001). *Les 5 niveaux d'appropriation des technologies de l'information et de la communication chez les enseignantes et les enseignants* (District scolaire no 1).
Parks, A., Pisapia, J., & Metropolitan Educational Research Consortium, R. V. A. (1994). *Developing exemplary technology-using teachers* (Research Brief #8).
Peters, M. (2015). Enseigner les stratégies de créacollage numérique pour éviter le plagiat au secondaire. *Revue canadienne de l'éducation, 38*(3), 1–28. http://www.cje-rce.ca/index.php/cje-rce
Peters, M. (2023). Note éditoriale: Intelligence artificielle et intégrité académique peuvent-elles faire bon ménage ? *Revue des sciences de l'éducation, 49*(1). https://doi.org/10.7202/1107846ar
Peters, M., Boies, T., & Morin, S. (2019). Teaching academic integrity in Quebec universities: Roles professors adopt [Original research]. *Frontiers in Education, 4*(99). https://doi.org/10.3389/feduc.2019.00099
Pierson, M. E. (1999). *Technology integration practice as a function of pedagogical expertise. Journal of Research on Computing in Education, 33*(4), 413–430. https://doi.org/10.1080/08886504.2001.10782325
Raby, C. (2004). *Analyse du cheminement qui a mené des enseignants du primaire à développer une utilisation exemplaire des technologies de l'information et de la communication (TIC) en classe.* [doctoral thesis, Université de Montréal]. https://theses.hal.science/file/index/docid/27773/filename/Rabythese.pdf
Solmon, M. A. (2018). Promoting academic integrity in the context of 21st century technology. *Kinesiology Review, 7*(4), 314–320. https://doi.org/10.1123/kr.2018-0042
Soppe, K. F. (2013). Student plagiarism: Intended or accidental? *Social Cosmos, 4*(1), 51–56.
Šprajc, P., Urh, M., Jerebic, J., Trivan, D., & Jereb, E. (2017). Reasons for plagiarism in higher education. *Organizacija, 50*(1), 33–45. https://doi.org/10.1515/orga-2017-0002
Thomas, M., Vandy, D., Kimberly, E., Jessica, E., Mary Helen, F., Emily, S., & Cristian, L. (2022). COVID-19 academic integrity violations and trends: A rapid review. *Education Sciences, 12*(12), 901. https://doi.org/10.3390/educsci12120901
Yeo, M. A. (2023). Academic integrity in the age of artificial intelligence (AI) authoring apps. *TESOL Journal, 14*(3).

Yu, H. (2023). Reflection on whether Chat GPT should be banned by academia from the perspective of education and teaching [Opinion]. *Frontiers in Psychology, 14*. https://doi.org/10.3389/fpsyg.2023.1181712

Zwick, M., Springer, M. L., Guerrero, J. K., DiVentura, D., & York, K. P. (2019). An activity to promote recognition of unintentional plagiarism in scientific writing in undergraduate biology courses. *Journal of Microbiology & Biology Education, 20*(2). https://doi.org/10.1128/jmbe.v20i2.1751

Chapter 6
Teaching Writing in the Time of ChatGPT: Rethinking What Counts as Learning

Alyson E. King and Pariss Garramone

Abstract From the perspective of undergraduate educators in the Social Sciences we ask: what are the challenges and opportunities associated with teaching writing in the time of ChatGPT, a powerful large language model (LLM) that generates written products and is impacting the way we communicate? With the advent of AI-driven LLMs, students now have access to sophisticated writing tools that generate text, and many are using them to complete written assignments. As a result, GenAI writing tools are helping to re-shape the conversation on teaching and learning in undergraduate education. Our chapter examines and traces the evolution of our thinking about undergraduate writing in our programs, our "way-finding" process, and the important pedagogical considerations that allow us to enact the curriculum in sensitive and meaningful ways that also build students' academic integrity.

Keywords Academic integrity · Generative AI · Writing · Critical thinking · Higher education · LLM

Introduction

The rapid development of generative artificial intelligence (GenAI) large language models (LLM),[1] such as Chat GPT, has reshaped the writing and learning habits of many students. These GenAI tools allow students to generate text without the need

[1] We are using the term GenAI in its popular usage, but we acknowledge that there is much discussion and debate over the terminology—especially regarding the term "intelligence"—and its trustworthiness. See, for example, Lewis et al., 2021; Lewis & Sarkadi, 2024; Lewis & Marsh, 2022.

A. E. King (✉) · P. Garramone
Ontario Tech University, Oshawa, ON, Canada
e-mail: alyson.king@ontariotechu.ca; pariss.garramone@ontariotechu.ca

A. E. King (ed.), *Artificial Intelligence, Pedagogy and Academic Integrity*, Advances in Artificial Intelligence in Education 1,
https://doi.org/10.1007/978-3-031-92534-4_6

for manual composition. As educators, it is crucial to understand how this technological advancement impacts the teaching and learning of writing skills, as well as critical thinking and analytical capacities. This chapter aims to investigate the challenges and opportunities posed by ChatGPT and develop effective curriculum and pedagogical approaches to using written assignments as a thinking tool in this time of GenAI advancement. Doing so requires thinking about what learning is, how the use (or non-use) of GenAI fits with learning outcomes, how these digital technologies have been, are, or could be used in classrooms, and what their use means for integrity in academic studies. Ultimately, anticipating what the future holds for learning in the context of GenAI is largely speculative and we aim to continue to collect data and examples to further explore the ways in which teaching and learning are changing.

The data for this project was collected as part of two projects. The first project is a large international partnership grant (Partnership on University Plagiarism Prevention (PUPP, PI Dr. Martine Peters, funded by the Social Science and Humanities Research Council of Canada). The second project (Teaching writing in the time of ChatGPT) is a small, unfunded project for faculty members to brainstorm together about how to manage the advent of easily accessible GenAI tools. Both projects were approved by the university's Research Ethics Board. The PUPP project has three stages of data collection with both instructors and students: an electronic survey (2023), individual (instructors) and group (students) interviews (2024), and a recorded writing task (2025). The data collection from the survey was completed in 2023 with 4661 students and 1357 instructors responding from universities in Canada, the United States, and Europe. While the bulk of this survey was composed of questions related to writing, informational, and referencing skills, two questions were about their use of GenAI. The second, much smaller project involved three one-hour discussions with colleagues in our university (located in Ontario, Canada), during which we explored our perceptions of how GenAI tools are impacting or may impact the way we teach undergraduate students. Our discussions took place during the summer as we prepared for the Fall 2023 academic term. After some planning meetings, one member of our group led a webinar introducing large language models (LLMs). This presentation allowed us to have a similar base understanding of the technology, how it works, and how it can be used. We had intermittent discussions as the academic year progressed and surveyed a random group of students (i.e., a link was disseminated to all students in our faculty) about their opinions and experiences of GenAI. While relatively few students responded, their answers were very similar to the responses from the larger survey. We (the authors) then reflected on the discussions with faculty members in conjunction with the quantitative data and existing literature.

Some of the most interesting and productive aspects of the conversation was the vulnerability displayed among group members in our meetings. Our discussions helped us to make sense, put into words and correct our own ideas of GenAIs and what we thought teaching might be like with the new technology. We discussed other technological changes, like the calculator, and tried to find similarities. The open discussion allowed for faculty members to situate themselves within the debates about using GenAIs in the classroom. One of our colleagues mentioned how she was going to connect with the hearts and minds of the students in her class. This idea of an embodied approach to discussing GenAIs has been important to the development of our ideas about our writing and research. Being open and recording our own sense making of learning about and with GenAIs is at the heart of this chapter. Our goal is to expose our process of understanding, learning and growing so that we can become better able to help students learn with and about GenAIs in our courses.

Surveying the Landscape

One of the first actions we took was to informally survey the landscape of what was happening at other universities and within Canada. Not surprisingly, since OpenAI's ChatGPT tools first gained popular attention towards the end of 2022, there have been many opinion and news pieces, reports, magazine articles, blog posts, workshops, seminars, and academic articles that explore a multitude of perspectives, from the dystopian to the utopian (e.g., Ascione, 2024; Baron, 2023b; G'Sell, 2024; Kelly et al., 2023; Lieberman, 2024; Mindzak & Eaton, 2021; Nafukho, 2023; Rutter & Mintz, 2023; Watson & Eaton, 2023; Zaphir & Hansen, 2024). Between Fall 2022 and Fall 2024, hundreds of GenAI tools were introduced that can generate writing, code, presentations, graphics, artwork, music, audio, and video (e.g., JotBot, Gemini/Bard, Tabnine, Beautiful.ai, Jasper.ai, DreamStudio, DALL-E 2, Magenta, LANDR, and Synthesia); even more have been developed since then (McGill, 2023; McKinsey, 2024). Effective regulations and guidelines for their development and use are slowly emerging at government and institutional levels. Many of these efforts are informed by the Montréal Declaration for a Responsible Development of Artificial Intelligence: "The Montréal Declaration is a collective work that aims to put AI development at the service of the well-being of all people, and to guide social change by developing recommendations with strong democratic legitimacy" (2018). Concerns about the privacy of users' information, the copyright and intellectual property rights of the training materials, and the quality of generated information have all sparked debate broadly; in educational settings, the debate has focussed on the impact on the learning environment and capacity of students to build critical thinking skills versus the need for students to be able to effectively and ethically use these tools in the workplace (Abdaljaleel et al., 2024; Moya et al., 2024; Pedersen, 2023; Salajan et al., 2024).

While the Canadian government introduced a voluntary code of conduct regarding the development and management of generative AI systems, at the time of writing no regulations or laws about its use have come into effect (Canada, 2023a). As a voluntary code, few companies have signed on, and no universities or colleges had done so by the end of 2023. This voluntary code is intended to "provide a bridge to regulations" that are currently proposed as the Artificial Intelligence and Data Act (AIDA). The Artificial Intelligence and Data Act (AIDA) was proposed as part of the Digital Charter Implementation Act (2022) and is intended to ensure that any AI systems used in Canada are designed, developed, and deployed in responsible ways and would hold businesses accountable for their development and use of AI systems (Canada, 2023b, c). As of mid-2024, the AIDA remains under committee consideration. Similarly, while many universities and colleges have created general guidance for students and faculty members, policies are still evolving as administrators, instructors, and others continue to grapple with anticipating the short- and long-term impacts of these tools. Research on the benefits and pitfalls in the use of AI chatbots and their impacts on learning is still emerging, but some positive outcomes have been seen (Wu & Yu, 2024). The AI Observatory website maintained by the Higher Education Strategy Associates (HESA) provides a regularly updated list of policies at Canadian and international universities and colleges, as well as other news and research about Artificial Intelligence as it relates to higher education (HESA, 2023). Other groups, such as the Partnership on Plagiarism Prevention (PUPP) and the Digital Life Institute (DLI), are researching and publishing about different aspects of GenAI in education. Clearly, the situation is evolving as we write, leaving students and instructors struggling to keep up with the rapid changes.

Authenticity in Writing: Engaging Hearts and Minds

As instructors who believe writing skills are connected to the building of capacity in critical thinking and analysis, we first looked to existing research on student writing in higher education and GenAI. Authenticity in writing assignments has been shown to be important for students of all ages; however, effective implementation of such assignments can be difficult without losing that sense of authenticity (Behizadeh, 2019). Research has demonstrated that authentic writing assignments tend to have flexibility built in whereby students can choose their topics and writing process; furthermore, a focus on social justice tends to increase engagement, strengthen a sense of audience, and help young people to navigate their world (Behizadeh, 2019). In an era of universities being pushed to be more career-related, Behizadeh (2014) argues that student writing tasks need to be more than "authentic to jobs" (cited in Behizadeh, 2019, p. 29). No matter one's perspective on the role of universities, engaging the "hearts and minds" (H. Scott, 2023, personal communication) of students is important.

Authentic writing prompts do more than improve student engagement; they make it more likely that students will learn and develop advanced skills. Soon after ChatGPT made front page news, the Association for Writing Across the Curriculum (AWAC) released a statement about the use of GenAI in curriculum explaining that writing is a fundamental mode of learning because understanding is developed via the writing process (AWAC, 2023). AWAC cautioned that "exclusively having AIs generate writing does not engage students in an essential mode of learning" (p. 2) and recommended that instructors encourage "the kind of deep learning and cognitive development that students gain through writing to learn and through learning to write in specific situations" (p. 2). The evidence for this idea of learning through writing is well-established (e.g., Harrison, 1983; Çavdar & Doe, 2012). For instance, Vasset et al. (2024) found that health science students benefited from written work that engaged both theoretical knowledge and clinical experience. However, this type of learning takes time and effort, something that will be increasingly challenging to require given the temptation of GenAI that allows tasks to be completed quickly and in a manner that likely meets minimum expectations.

The process of writing as a way of developing critical thinking and engaged learning has been linked to broad educational developments in the fields of neuroscience and psychology (Berninger et al., 2006; James & Engelhardt, 2012; Mueller & Oppenheimer, 2014). For example, taking written notes as one learns about a topic and translating those ideas into one's own words is an important step in the process of understanding that topic. This note-making step to understanding is centered on using writing to capture and translate the process of coming to know. Similarly, Fraser and Spiller (1996) describe the importance of developing writing and thinking skills within disciplines (rather than as an isolated writing course): "learning to write articulately about a subject cannot be separated from the acquisition of knowledge about the subject itself" (p. 93). In other words, it is necessary to be able to fluently express one's understanding of a subject or issue. The process includes everything from the brainstorming phase, discussing ideas, writing, and peer editing, as well as being able to provide feedback to their peers (Fraser & Spiller, 1996). The impact on long-term memory of cognitive offloading by relying on digital devices to quickly look up information or use a GPS navigation guide appears to be significant. Cognitive offloading occurs when digital technologies are used externalize knowledge, undermining human capacity for remembering what has been looked up or photographed. Studies of people using GPS for navigation, for instance, have demonstrated negative impacts on brain elasticity, hippocampus volume and ability to navigate without relying on a digital tool (Baron, 2021). With regard to the impact of reading online texts, there is considerable concern that "quick, shallow reading … undermines engagement with deeper analytical thinking" (Baron, 2021, p. 33; see also, Baron, 2017). This research, we believe, flags some serious consequences if reliance on new tools like GenAI becomes more prevalent.

Worries about the sustained reading ability of undergraduate students has been of concern for over two decades. Anecdotally, students in my undergraduate courses often tell me [Alyson] that the last time they read a book for pleasure (and not as an assignment) was in about grade 10. Given that evidence shows that students' reading has become fragmented (Liu et al., 2022) and that "[s]ustained reading is on the decline, including in educational contexts," the advent of GenAI and LLMs raises concerns about the long-term literacy of populations (Baron, 2021, p. 35). In an effort to engage students better, some instructors have reduced "both the amount and complexity of undergraduate reading" being assigned and increased "use of audio or video assignments (e.g. podcasts, TED Talks) [instead of] text-based materials" (cited in Baron, 2021, p. 35). Other researchers identify the importance of having students "develop, seek out, and adapt a variety of literacies, or what have come to be called … 'multiliteracies'" (Alexander et al., 2020, p. 105). As a result, they argue, writing skills are taking on even more importance as "a site of intellectual, moral, and civic development" (Brandt, 2014, cited in Alexander et al., 2020, p. 105). Nonetheless, multiliteracies are important (Hughes & King, 2010; Hughes et al., 2011) given the changing educational landscapes and need to accommodate students in different ways. As a result, instructors need to be intentional and thoughtful about their pedagogical approaches, whether digital literacies, multiliteracies or traditional literacies are used.

As types and modes of professional writing have burgeoned (e.g., blogs, podcasts, and websites, in addition to longstanding forms such as marketing and advertising, news, professional reports and articles), expectations have changed about what students now need to learn, including how to write well in different forms and use emerging tools. Alexander et al. (2020) use the metaphor of "wayfinding" to describe efforts to navigate the different forms of writing that authors, including students, must manage. Multiliteracies, especially digital literacies, are important here because a range of literacies can be deployed for different purposes including activism and political engagement, identity development, and learning. In some areas of writing studies, researchers acknowledge that the process of thinking and writing involves much more than simply knowledge transfer and proficiency; writing involves "a significant, transformative shift in how one understands … knowledge" (Alexander et al., 2020, p. 117). The idea of wayfinding in writing and literacy, as described by Alexander et al. (2020), is "an attempt to develop a way of thinking about writing and literacy that would place an emphasis on the complex and recursive movement in and out of different territories, realms, spaces, and spheres of writing ecologies" (p. 121). It will be necessary for students to be able to navigate these territories as members of the workforce.

How technologies are discussed is important here because metaphoric terminology tends to shape how a technology or an object of study is perceived and understood (Alexander et al., 2020). Take, for instance, the critique of GenAI references to its "hallucinations." The manner in which GenAI is spoken about affects how it is viewed in different sectors of society and, in turn, can shape the way the world is viewed (Nascimento & Chown, 2023). While such metaphors can help one to understand new technologies, they may also play a more insidious role in paving the path towards acceptance of technologies that may have unwelcome and unanticipated consequences. Metaphors, such as the use of "hallucinations" to describe GenAI's tendency to fabricate or falsify ideas and information, personifies the technology which in turn blurs its actual nature. Similarly, the idea that software, algorithms or machines are "intelligent" (as in Artificial Intelligence) encourages users to unquestioningly rely on the output of GenAI software without critically examining it. In this regard, the more GenAI tools are described as being thoughtful or creative entities, the more likely students and others will be inclined to take GenAI's output at face value without questioning its "expertise" (i.e., training sources) and reliability.

Baron's (2023a) comments on how students perceive the difference between reading print versus digital texts are interesting. Digital texts are perceived as being easier and faster to read, less boring and more engaging; at the same time, however, evidence suggests that retention of information and comprehension of materials read digitally is much poorer. Furthermore, research in the field of Rhetorical Genre Studies demonstrates that genres (i.e., a type of text such as poetry in literature, or types of film, video games, music and so on) can be seen as social action and "flexible communication strategies" (Artemeva, 2004, pp. 3–4). In this regard, the process of writing is as important as the final product, both of which are influenced by the writer's social location in their day-to-day lives. The use of GenAI in writing tends to create generic products because "LLMs do not engage in the full human activity of writing," even though they may simulate genre conventions (Omizo & Hart-Davidson, 2024, p. 273). Furthermore, Vee (2023) notes that GenAI is never uncertain; yet, uncertainty is something that students experience as they explore and seek to understand new ideas through their research and writing. If writing is simply a way to communicate information, GenAI may help; if writing is for thinking and learning by wrestling with new and challenging ideas and situations for understanding, then instructors must create opportunities for students to write in different ways. Some of the current research from Schnitzler et al. (2024) includes the development of the TextGenEd: Continuing Experiments repository that collects and shares open-access assignments to help educators engage with GenAI in ways that nurture an embodied process to learning and writing. As they state:

> teaching is not about efficient solutions. We want students to do it right, where *right* is an expansive, ever-shifting, human goal connected to their own desires, skills, and the world beyond the classroom. We want writers to use their embodied, emotive selves and experiences to inform the writing and thinking that they do. And we want the technology that they use in their writing to support, not supplant, their humanity. (2024, para. 4)

Is It Cheating to Use GenAI?

Whether or not the use of GenAI in academic work is academic misconduct has become a frequent point of conversation within higher education. The short answer is "it depends." Most academic integrity policies contain statements about students only submitting for grading their own work completed only by themselves. Anything else—buying an essay, submitting an assignment done by someone else, and so on—is considered academic misconduct. Rogerson and McCarthy (2017) describe the use of online paraphrasing tools to include in one's own written work as "facilitated plagiarism," something that undermines students' understanding of the paraphrased material. Not only is it important to attribute, cite, and reference all information and knowledge created to avoid accusations of plagiarism, it is also necessary to find ways to ensure that students really are demonstrating their understanding of their work. Sarah Eaton at the University of Calgary has written extensively about issues of academic integrity. In a blog post about what she calls post-plagiarism, Eaton (2023) notes that it will become normal to have writing co-created by humans and GenAI, a form of "hybrid writing." In her view, this hybrid writing will enhance creativity and help language barriers disappear. At the same time, Eaton argues, human writers will retain responsibility for fact-checking and verifying what has been written. Given that large proportions of students in higher education (Chan, 2023) may be using tools like ChatGPT or Grammarly to complete assignments and help with homework, norms regarding the use (or non-use) of GenAI must be forged. Furthermore, forms of generative and predictive AI are already integrated into many areas of research, industry, business, and the economy generally (Chan, 2023) and require that people graduating from higher education have some level of familiarity with them. Nonetheless, the nature of plagiarism as being surface rather than deep learning (Peters et al., 2019) mirrors the view that the uncritical use of AI/GenAI without deep thinking tends towards surface learning.

The amount of research on academic integrity is growing. In their research examining the security of online assessments, for example, Garg and Goel (2022) found that technological tools are used by students for five types of dishonest behaviour, as well as by instructors for detection and prevention. They developed a concept map that describes mitigation strategies (related to assessment format, integrity policy, computer lockdown, authentication tools, authorship tools, proctoring, and data analysis) for five types of dishonesty (impersonation, forbidden aids, collusion, plagiarism, and gaming the system). Furthermore, they provide an overview of the different roles that students, instructors, institutions, and platform providers play in preventing and detecting academic dishonesty. This article is interesting because it collates a range of articles about different integrity strategies for the five types of dishonest behaviours, many of which can be modified to suit this new era of GenAI in academia.

Other researchers have pointed to the influence of technology on the likelihood of plagiarism occurring and the fact that those who plagiarise will only attain a surface level learning of course content (Peters et al., 2019). By framing plagiarism as poor writing skills rather than as a moral issue, instructors can focus on teaching students strong academic writing skills in their disciplines and thereby prevent much plagiarism, especially the accidental instances of missing or poor referencing and the like (Adam et al., 2017; Jamieson & Howard, 2019; Peters et al., 2019). We highlight this point because we argue that the advent of GenAI and LLMs aimed at content creators, especially for written work, will have a significant impact on students' abilities to write coherently and with integrity. Indeed, we fear that without intentional instruction about using writing to develop critical thinking and communication skills, graduating students will be placed at a disadvantage if they have relied too much on using GenAI tools for their written work. At the same time, we acknowledge that students will need to know how to use these tools effectively in the workplace.

As Adam et al. (2017, p. 30) argue, "writing needs to be viewed as a process, or something that can be constantly improved." We believe that a similar approach needs to be taken in the face of increasing access to GenAIs. Since informational skills are cross-disciplinary (Peters et al., 2019), instructors in all disciplines need to pay attention to how students are accessing information for their assignments, processing it, and thinking critically about its meanings. The question of instructor workload also plays a role in these discussions. Indeed, some instructors tend to reduce their expectations of students in order to reduce their own workloads: active learning, scaffolded written assignments, and longer, frequent assessments can translate into a higher workload for instructors (Kuh, 2003). At the same time, as hooks (2010) argues, "[c]hoosing to maintain high standards for pedagogical engagement and performance is one way to ensure that integrity will prevail" (p. 32). Building a course that engages students such that they are able to develop strong writing capacity, critical thinking skills, and knowledge about how to use the GenAI/LLM tools effectively and ethically will require instructors to be intentional and creative. At a basic level, we can build on hooks (2010) comment that "[i]ntegrity is present when there is congruence or agreement between what we think, say, and do" (hooks, 2010, p. 32), to argue that writing that is not authentic is lacking in integrity.

Decolonial, Authentic Writing Pedagogy

The current literature on decolonizing higher education has focused on the need to dismantle colonial mechanisms that keep long-standing patterns of power and knowledge relations in place (Zembylas, 2021). Curricular content, course design and the use of digital technologies within course modules are spaces that tend to reproduce colonial formations of the past. In particular, algorithmic coloniality (Mohamed et al., 2020) is one manifestation where the use of GenAI technologies deploys biased, Western-centric algorithms that privilege whiteness. From the use of search engines to GenAIs, algorithmic coloniality structures and controls the information used in digital educational spaces (Noble, 2018).

McKnight and Shipp (2024) help to unpack the metaphors used to describe and explain GenAI's role in teaching writing through their critical examination of the concept of "tools," and in particular, "colonial tool thinking" (p. 27). By interrogating how the relationship with GenAI is being constructed through language, McKnight and Shipp expose the colonial relations being reproduced. They historicize how "tools" are often a category through which to view, for example, Indigenous communities and note that many colonial museum collections showcase handcrafted native "tools" isolated from their deeply embedded social meaning and relation to community. McKnight and Shipp remind us that tools shape us, and that the relations often hidden in the term "tools" include "vast systems of extractions and logistics, with supply chains that wrap around the entire planet" (Crawford, 2021, as cited in McKnight & Shipp, 2024, p. 29).

Zembylas (2021, p. 33–35) suggests three strategies for the development of a decolonial ethical framework for GenAI in higher education. These three strategies include: (1) historicizing AI and digital technologies to uncover hidden assumptions in order to promote algorithmic transparency; (2) nurturing practices of ethical solidarity so that communities can heal from coloniality; and (3) creating renewed affective and political communities that are based on embodied experiences of connection and friendship. These three proposed strategies focus on "the importance of the visceral elements of decolonization, the embodied experiences of love, solidarity, and friendship" (p. 35). Similarly, for Khanna (2020), undoing the affective violence of colonialism requires healing and building human relations. One approach, culturally digitized pedagogy (CDP), "distinguishes itself by concentrating on the utilization of digital tools and platforms to enhance critical awareness concerning youth activism, advocacy, and allyship in the context of the social justice issues they take on" (McDaniel, 2024, p. 6). This type of intentional use of digital tools and platforms can aid in thinking about how to effectively and ethically use GenAI.

Research on reading with digital texts versus print (e.g., paper-based books or newspapers) demonstrates that students tend to remember better with print texts, especially for longer texts, when more detail, inference or abstract thinking are needed in answers to questions (Baron, 2017, 2021; Hargreaves et al., 2022; Singer & Alexander, 2017). Similarly, students tend to do more poorly on assessments in digital formats than in print formats (Baron, 2021, 2023a). Taking these findings along with bell hooks' argument about the importance of a decolonial approach to teaching and learning (hooks, 2010), we highlight the need to look beyond the simple view that GenAI is a tool that we can use for our own purposes. Acknowledging that artificial intelligence evolved out of the generally colonial, white, male-dominated field of computer science and gaming (Johnson & Pietri, 2023; Perez-Felkner et al., 2012) requires instructors to look at how students from all backgrounds can learn about both GenAI and writing skills in ways that will benefit them throughout their life course rather than simply reinforcing social biases and racism.

Perceptions of GenAI in Higher Education

Across the higher education landscape in Canada and around the world, universities, colleges, faculties, instructors, and students have differing opinions about and approaches to managing the use of artificial intelligence (AI) tools, such as GenAI, paraphrase generators, and LLMs, in academic activities. For instance, in one case, students are told "the use of AI to complete an academic evaluation [is] a violation of the Academic Integrity Policy unless a professor has specifically authorized the use of the AI in question" (Miedema, 2023, p. 2) and guidance is provided to instructors regarding when and why such tools may be used. Other universities have provided guidance on wording for syllabi so that it is clear to students what the expectations are. This approach places the responsibility on instructors for decisions about whether or not to employ GenAI in their classes based on a course's learning objectives. This premise makes sense since instructors have the academic freedom to teach their courses in the pedagogical and andragogical manner that they think is best for meeting course learning objectives. At the same time, it also increases workload; instructors may feel overwhelmed by the vast amount of hype, conflicting information and viewpoints about GenAI and burdened by the need to learn how to talk knowledgeably about GenAI in the classroom (Pedersen, 2024).

In the 2023 survey conducted by the Partnership for the Prevention of Plagiarism (PUPP) not long after the launch of ChatGPT, over half of the undergraduate students at our university indicated that they had never or rarely used an artificial intelligence tool to help with their writing assignments (n = 199 of 361 respondents) while 67 stated they sometimes, frequently or always used a tool (95 did not answer the question). Although the numbers were low (the surveys were administered only a few months after the launch of ChatGPT) for all fields of study, most of the respondents who indicated that they used a GenAI tool in their written assignments were in business, computer science, engineering, or the sciences. For instance, 50% (n = 14) of 28 computer science students, 26% (n = 14) of 54 Business students, and 31% (n = 13) of 42 engineering students had used a GenAI tool to complete assignments. By comparison, 15% (n = 6) of 40 social science students and 22% (n = 4) of 18 education students had used an GenAI tool to complete assignments. In the international (Canada, the US, and Europe) survey of students, 62% of respondents (n = 2219) said that they never used a GenAI tool to complete their assignments, while 14.5% (n = 524) said that they rarely did so. By comparison, 14% (n = 501) sometimes, 6.8% (n = 246) frequently, and 2.8% (n = 102) always used these tools. Other researchers using different methods have found that students tend to feel "a high level of comfort in adopting new technology" like ChatGPT and are influenced by high expectations about performance (Strzelecki, 2023, n.p.). It is likely that as students continue to become more comfortable with the technology, and as it improves, more students will be inclined to experiment with and use the tool in all facets of their lives (Tables 6.1 and 6.2).

In our other, smaller pilot survey conducted in 2024, most students (n = 11 of 12 respondents) indicated that in all or some of their classes their professors had discussed GenAI. Of these students, only one responded that the use of GenAI is not cheating; all others indicated that it was or might be cheating. In written comments, students' attitudes about GenAI and cheating reflected the range of opinions represented in popular media reports and blogs.

Table 6.1 Have you used an AI tool to help with your assignment?

	Local student survey		International survey	
Never or rarely	199	75%	2743	76%
Sometimes, frequently, or always	67	25%	849	24%
	266		3592	

Table 6.2 Have you used AI tools in your assignments by field of study (Local student survey)

Computer Science	14 (of 28 students)	50%
Business	14 (of 54 students)	26%
Engineering	13 (of 42 students)	31%
Social Science	6 (of 40 students)	15%
Education	4 (of 18 students)	22%

- "I believe that the direction is given to the AI by the student but the content is pulled from others. There could be circumstances to have AI do the work I suppose, but I will say, as an employer that hires, cover letters that I believe to be AI-generated often are not reviewed by the individual and it is evident, and I am not sure who I am really interviewing until I meet them so I tend to pass on those applicants."
- "G[en]AI can be used to cheat, if it is not used correctly for example if the student just purely copy and pastes. However, if students use G[en]AI responsibl[y] to just brainstorm ideas, fix grammar, find other ways to form their writing for when they can't find ways to say the information. As long as the student put[s] in their own work and only uses G[en]AI as a tool to make their own work [better], it is not cheating."
- "Depends on how it's used, it's not cheating if it's used to help format work or spell/grammar check."
- "In your written paper it is cheating, but using it for presentations, getting ideas to say a speech, brainstorm for a topic, collect source evidences quicker, make easy notes from a reading with concepts is beneficial."
- "If the student is using AI to write their assignments for them or is incapable of demonstrating their own learning because they have been using AI to complete their work, then it could be considered cheating. Using AI for brainstorming, generating ideas, creating comprehensive study notes or as a tool for studying should not be considered cheating."

These students illustrate the nuances and complexities of using GenAI. At least one student commented on the possibility of GenAI:

- "AI cannot totally replace humans, however if we don't use skills we lose them. Our skills will definitely change in [the] future. AI could also help guide people that need information to get them through certain situations. It is a great option but we also have to be able to discern what information is accurate and that is getting more difficult to navigate, no fault of AI, I think!"

This student raises an important point regarding the potential loss of skills if GenAI becomes relied upon for essential skills. In general, student perceptions of GenAI and related tools reflect popular opinions about GenAI, both positive and negative. These views include concerns about the protection of privacy, increasing job losses to AI, added opportunities for cheating, and even the potential to cause "the downfall of humanity." On the positive side, students see benefits in reducing human error, assisting in brainstorming ideas, and helping to improve their own work.

Some instructors have long used tools such as text-matching software that do the work of flagging potential instances of copying and pasting in student essays. In the local PUPP survey, 11 out of 18 instructors responded that they frequently or always use text-matching software (e.g., Turnitin, Urkund, Compilatio). Yet, among the international respondents, over a third (n = 510) of instructors said they never or rarely used text-matching software and only 16.5% (n = 224) said they always or almost always used such software. The regional differences in the use of text-matching software and other digital tools could provide an interesting analysis in future research. As GenAI became more accessible, these text-matching systems attempted to integrate AI detection capacity. At the time of writing, these systems have limited use in effectively and accurately detecting the use of GenAI in written work. Déri explores the place of text-matching software further in this volume.

Critical Thinking

bell hooks (2010) argues that

> [t]hinking is an action. For all aspiring intellectuals, thoughts are the laboratory where one goes to pose questions and find answers, and the place where visions of theory and praxis come together. The heartbeat of critical thinking is the longing to know—to understand how life works. Children are organically predisposed to be critical thinkers. (p. 7)

Indeed, critical thinking includes the capacity to reason and to reflect critically on one's beliefs and actions to see that there are better (e.g., seeing both sides of an issue) and worse (e.g., biased or based in fallacies) forms of reasoning; it is about moving beyond the simple acquisition of information and content knowledge to identifying and challenging existing assumptions and finding alternative ways of seeing and acting in the world (Bean, 2011; Bean & Melzer, 2021; Ellerton et al., 2022). In academia, teaching critical thinking is about helping students see the problems that surround them in their daily life; whether from the personal and local (micro) to whole world (macro) perspectives, awareness and understanding of invisible and normalized inequities and paradoxes are key. As Bain (2004) suggests, "natural critical learning environments" are all around students and educators can help students to see these "intriguing, beautiful, or important problems, [as] authentic tasks that will challenge them to grapple with ideas, rethink their assumptions, and examine their mental models of reality" (p. 18). In this approach, a critical thinker is someone who can see different sides of an issue and is open to changing one's mind about their ideas and views based on evidence from facts and problem-solving. At the same time, a critical thinker must be able to justify and evaluate those ideas and beliefs by drawing inferences and conclusions from existing research (Ellerton et al., 2022).

Many researchers, such as Bean (2011), see writing as a key part of "the process of doing critical thinking" (p. 4); the written artifact is the result of that process. With increasing access to GenAI and related tools, some researchers are pointing to the idea that the structure of education systems and the roles of teachers will shift to being more about developing self-directed and individuated critical thinking rather than providing information and being skill focussed; in other words, there will be a more student-centred approach to using the tools of cloud computing, artificial intelligence and big data (Wei, 2021). Others are exploring ways to teach critical thinking such that students are focused less on knowing something and more on the thinking process (Ellerton et al., 2022). Given the perception that GenAI potentially "robs students of motivation to write and think for themselves" (Baron, 2023b, p. 1), it is urgent for university instructors to find ways to foster the critical thinking capacities of students and opportunities for them to use those skills.

Tools and Strategies

Ellerton et al. (2022) provides some interesting tools for teaching critical thinking in high school settings; these can be used in university teaching, especially in the context of easy access to GenAI. For instructors, Ellerton et al. (2022, p. 3) suggest starting by asking three questions to help plan in-class thinking, starting with how do I know students are thinking in my classroom? We argue that these questions should be in a slightly different order so that instructors start with planning for student thinking, followed by asking how do I know that thinking is occurring and, finally, giving feedback to students on the quality of their thinking:

1. How do I plan for thinking to occur?
2. How do I know students are thinking in my classroom?
3. How do I give students feedback on the quality of their thinking?

Underlying these strategies is the basic premise that students learn by doing. In other words, students will learn the skills of critical thinking by engaging in it. Ellerton et al. (2022, p. 6) explicate a useful tool for visualizing basic critical thinking actions, a tetrad that illustrates the cognitive skills involved with critical thinking. In light of GenAI, we ask: Does this tetrad need to be revised to accommodate GenAI use? GenAI could be used at any of these critical thinking touch points, but should it? Since we want students to move beyond, for example, basic inference thinking using only their personal experiences and common assumptions (e.g., hearsay, myths, ideology, misinformation, conspiracy theories), we also ask, can GenAI help to encourage students to move beyond basic inference and interpretation (i.e., inferring or interpreting based on "common sense" assumptions or common occurrences)?

Building on Ellerton et al.'s (2022) tetrad of cognitive skills, instructors can consider how GenAI might support the building of these skills. While GenAI may appear to engage in the other components of critical thinking, its nature limits its capacity for deep learning based on analysis, evaluation, inferring and interpretation. Furthermore, students must engage in critical thinking to be able to coherently justify their interpretations and conclusions. Figure 6.1 illustrates the places (Brainstorm & Organise; Explain; Hypothesize) where GenAI might be most useful in supporting student work and where students should be encouraged to develop their critical thinking skills (Analyse & Evaluate; Interpret & Justify; Infer).

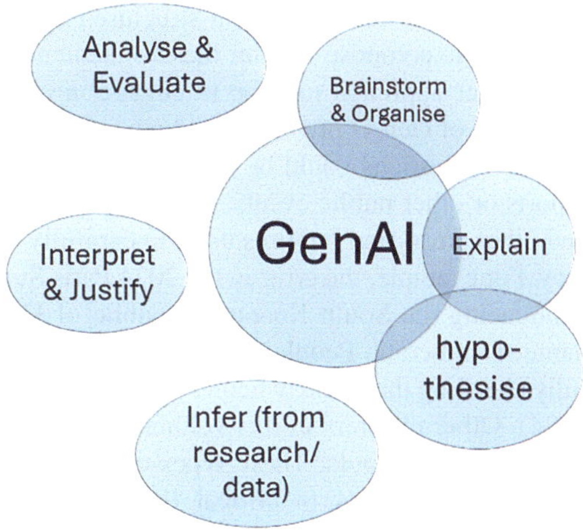

Fig. 6.1 Integrating GenAI with cognitive skills for critical thinking

Planning for Thinking

Starting with the first question, how do I plan for thinking to occur, we propose building on research by Bean (2011) about the principles for critical thinking and writing. Drawing on the eight principles to support critical thinking developed by Kurfiss (1988 cited in Bean, 2011), Bean (2011) highlights the importance of three of those principles for critical thinking and writing. Here Bean (2011) advocates for using (1) "problems, questions, or issues [as] the point of entry into [a] subject" (p. 5); and (2) using carefully designed and relevant questions and issues to create "a source of motivation for sustained inquiry" (Bean, 2011, p. 5). They further argue that (3) shifting the focus of courses from being "lecture-centred" to being "assignment-centred" can allow course goals and evaluations to "emphasize using content rather than simply acquiring it" (p. 5). In this approach, assessments and activities can be designed so that "[s]tudents are required to formulate and justify their ideas in writing or other appropriate modes" (Bean, 2011, p. 5). While these recommendations were made before GenAI became pervasive and are generally elements of good teaching practice, it is important to reflect on what this means in the age of GenAI.

Using GenAI to develop and promote critical thinking can occur with careful planning. By working with students to create prompts and build a strategy of question and answer, it is possible to encourage students to delve deeply into a subject. In the same way that an instructor might use a dialogic method in the classroom rather than lecturing, a similar approach can be used with GenAI tools. For example, a student who has come up with a thesis statement for an essay could ask a GenAI tool: Can you help me with my thesis? Does my thesis answer a question? Furthermore, a student could ask: Does my thesis make a claim that can be supported by evidence? In this way, thesis statements may be improved by engaging in a dialogue that demonstrates a structured approach. By doing this, students may learn how to compose a better thesis statement and understand why.

Another approach may be to engage in class discussions centered around the reliability of GenAI products and how to fact-check the information generated. For instance, questions could be asked about where students have seen errors in news reports or other public events. GenAI has a reputation for creating misinformation and other errors and requires users to carefully reflect on and confirmation of accuracy. For example, the error in the 2024 Paris Summer Olympics opening ceremony announcing the South Korean (Republic of Korea) team using the North Korean name (Democratic People's Republic of Korea) could have been avoided by carefully checking the name, whether or not it was the result of GenAI use (Wertheimer, 2024). Other chapters in this volume demonstrate the unreliability of current versions of GenAI tools. These types of errors can serve as jumping-off points for introducing students to critical thinking via fact-checking and meta-cognitive thinking.

One form of meta-cognitive thinking using GenAI is what one of our colleagues calls The 5 Whys (M. Shane, 2023, 5 whys in-class activity, personal communication). In the classroom, rather than lecturing, the instructor poses a "why" question to GenAI about the subject matter for the course. To go deeper into each question, students must come up with additional "why" questions that probe deeply into the content. In political science, for instance, one could ask:

> ***Why does democracy matter in the 21st century?***
>
> *Edge Co-Pilot Response:* "Democracy remains a cornerstone of political organization in the 21st century, as it embodies the principle of popular sovereignty, where the government is created by, for, and of the people. This system of governance, particularly in its liberal form, ensures that individual freedoms and rights are protected by a constitution, which is crucial for maintaining a just and free society. Despite challenges such as political polarization and the rise of illiberal democracies, the democratic process promotes progress through peaceful competition between different interests and ideas. It provides a framework for citizens to engage in their communities, counter threats to self-government, and rebuild trust in political institutions. As the world continues to evolve with technological, demographic, and cultural changes, democracy adapts, offering a resilient model that can address the complexities of modern societies."

In the class, discussion could include prompts such as: What does this mean in the day-to-day world? What examples do you see around you? What are the key points in the Co-Pilot response? Find peer-reviewed references using the university library databases to support this response; how accurately are they represented in the response? What are the next "why" questions to ask to go deeper? While there are many other questions that could be asked, the goal is to encourage students to ask questions about what they are learning overall. A further step might be to assign readings that address the questions posed and then compare with and critique GenAI outputs.

For instructors, being faced with the conundrum of how to teach something that we are still learning to use is troubling. How do we instructors learn something new that is outside of our field and then use it immediately in our teaching? Ellerton et al.'s (2022) first of their "pedagogical imperatives" might fit here: "1. Shift the focus in the classroom from knowledge to inquiry" (p. 8). Using this model, an instructor could explore the use of GenAI alongside the students using some of the strategies described above. While many instructors hesitate to demonstrate vulnerability or a lack of knowledge, students often feel more engaged when they are learning with their instructors; it is an excellent way to model critical thinking and research skills. And it opens an opportunity to learn from students, which may build their understanding and self-confidence.

Identifying Student Critical Thinking

Many of the suggestions we have included here approach learning as a conversational process (Bakhtin, 1981; Bean, 2011; Buber, 1947; Freire, 1970):

> We converse with ourselves to reflect on our current knowledge and question assumptions, and we converse with others to reach mutual understanding. A conversational learning system is one that connects conversational agents in a continual process of interaction to explore differences, gain experiences, and reach agreements. (Sharples, 2023, p. 161)

Sharples (2023) builds on the work of Pask (1975, cited in Sharples, 2023) who foresaw the potential for artificial intelligence to be conversational agents in "educational dialogues" (Sharples, 2023, p. 161). Given that collaborative and dialogic learning are accepted methods in higher education, a well-designed GenAI tool could support this form of social learning. For those who prefer not to interact or work with other students for reasons of personality, time poverty, or preference, GenAI could provide an alternative. However, in a world where students are reported to be struggling with loneliness and lack of engagement with their peers, are universities adequately supporting students in their studies and preparing them for their future if technology is overly relied upon? While this type of GenAI tool may work well as a backup or safety net, we argue that a better approach is to build social and conversational opportunities into classroom work (both in-person and online) so that students can strengthen their critical thinking skills while engaging with fellow students. Indeed, Ellerton et al.'s (2022) third pedagogical imperative is: "Work collaboratively where thinking can be shared" (p. 8).

As exploration of GenAI occurs, we believe instructors need to consider how to convince students that they need to be able to think, talk and write about a reading or course content without relying on a chatbot. In some ways, there is an inherent paradox—how can instructors convince students that they need to do the hard work of "thinking" and "writing" when there is an easy tool to use that appears to do that for them? One approach may be to use a coaching approach to the writing process. As Bean (2011) suggests, fostering class discussion and student participation to engage and develop a class assignment can work well. These group discussions can focus on generating ideas, clarifying grading with group norming sessions, or using model assignments to help students understand feedback (Bean, 2011, pp. 303–304).

Feedback to Students

This coaching model uses dialogue to help students talk about their thinking and writing, which in turn allows students to understand the requirements of an assignment and so that students are better prepared to peer review other students' work in the class. Peer review, in particular response-centred peer review (Bean, 2011, p. 298), where small groups of students respond to each other's written drafts, provides students with opportunities to not only share their thinking and writing, but to also learn how to effectively provide supportive feedback to their classmates.

In terms of feedback provided for students on their work, Bean (2011) reminds us that, "the best kind of commentary enhances the writer's feeling of dignity" (p. 317). Using feedback as part of a coaching approach can support meaningful learning. Zull (2002) has shown that positive emotions enhance cognition and comments that help students feel confident and hopeful allow students to have more engagement in a task (p. 61). To help develop impactful feedback for students, Bean recommends one-to-one conversations with students, as well as written comments in the style of personal correspondence with a supportive tone and focus on providing students with positive messages to encourage revision and improvement (p. 336).

We have discussed GenAIs in terms of "tools" for helping students write assignments; yet, it is also important to remember that GenAIs are now playing a larger role in the grading and management of student work. Ellerton et al.'s (2022) question, "How do I give students feedback on the quality of their thinking?" (p. 3) clearly points to the goal of using assessments to measure learning and to improve critical thinking. In their question, Ellerton et al. (2022) also re-frame the importance of feedback by not attaching it to assignments or to writing, which helps to expand what we might want to use as evidence of learning and artifacts of thinking in our classes.

The idea of writing to learn (Zinsser, 1988) has been central to many writing classrooms. Using writing tasks to record thinking and then spending hours providing comments on written work has been a foundation. What if GenAIs can help us think differently about thinking, about the ways we show thinking, and the ways we do thinking in our classes? How might GenAIs help (maybe even push us) to address Ellerton et al.'s (2022) questions by finding new ways to document and show thinking? What might happen if assignments are dialogic, discussion-based, model-based, and collaborative? What might happen if the feedback process is also dialogic, collaborative, discussion oriented, and model-based?

Conclusion and Next Steps

Throughout the chapter, we have discussed how the landscape of teaching and learning in higher education has been dramatically altered by the development and use of GenAI. This change to writing as a tool for learning provokes not only new pedagogical interventions, but a recognition that not just what we teach will change fundamentally, but also how students learn and the pace of that learning. The words of bell hooks (2010) are particularly useful here, as she argues,

> The vital link between critical thinking and practical wisdom is the insistence on the interdependent nature of theory and fact coupled with the awareness that knowledge cannot be separated from experience. And ultimately there is the awareness that knowledge rooted in experience shapes what we value and as a consequence how we know what we know as well as how we use what we know. (p. 185)

From hooks, the experience of writing allows students to apply what they are learning by testing out ideas and making mistakes rather than just "getting it right" which is often the outcome when students use a chat bot to complete their assignments.

As we look at extending our research and continuing our discussions with colleagues, we are particularly aware that we need to pay attention to (a) the long term effects of an education system that relies upon virtual resources and information and (b) how educators can mitigate the problems and better prepare students for a world reliant on digital technologies, while also nurturing hands-on experiences emphasizing "critical thinking and practical wisdom" (hooks, 2010) in a digital world. At the time of writing, we were encouraged to see how some educators were actively designing and using course-specific chat bots to facilitate more individualized feedback and support for students (see the work of Carter Zenke). In other words, it is not enough to simply throw up our collective hands and say we must go back to print-based texts; we need to explore how to work within the new normal to ensure young people can develop both retention capacity and the skills to navigate and exploit digital technologies effectively.

References

Abdaljaleel, M., Barakat, M., Alsanafi, M., Salim, N. A., Abazid, H., Malaeb, D., Mohammed, A. H., Hassan, B. A. R., Wayyes, A. M., Farhan, S. S., Khatib, S. E., Rahal, M., Sahban, A., Abdelaziz, D. H., Mansour, N. O., AlZayer, R., Khalil, R., Fekih-Romdhane, F., Hallit, R., … Sallam, M. (2024). A multinational study on the factors influencing university students' attitudes and usage of ChatGPT. *Scientific Reports, 14*(1), 1983. https://doi.org/10.1038/s41598-024-52549-8

Adam, L., Anderson, V., & Spronken-Smith, R. (2017). 'It's not fair': Policy discourses and students' understandings of plagiarism in a New Zealand university. *Higher Education, 74*(1), 17–32. https://doi.org/10.1007/s10734-016-0025-9

Alexander, J., Lunsford, K., & Whithaus, C. (2020). Toward wayfinding: A metaphor for understanding writing experiences. *Written Communication, 37*(1), 104–131. https://doi.org/10.1177/0741088319882325

Artemeva, N. (2004). Key concepts in rhetorical genre studies: An overview. *Discourse and Writing/Rédactologie, 20*(1), 3–38. https://doi.org/10.31468/cjsdwr.524

Ascione, L. (2024, February 20). The AI effect: How AI fuels growth and crafts job-ready grads. *eCampus News*. https://www.ecampusnews.com/teaching-learning/2024/02/20/the-ai-effect-higher-ed/

Association for Writing Across the Curriculum (AWAC). Statement on artificial intelligence writing tools in writing across the curriculum settings (January 2023). https://wacassociation.org/statement-on-ai-writing-tools-in-wac/. Accessed 30 July 2024.

Bain, K. (2004). *What the best college teachers do*. Harvard University Press.

Bakhtin, M. (1981). *The dialogic imagination* (C. Emerson & M. Holquist, Trans.). University of Texas Press.

Baron, N. S. (2017). Reading in a digital age. *Phi Delta Kappan, 99*(2), 15–15.

Baron, N. S. (2021). Know what? How digital technologies undermine learning and remembering. *Journal of Pragmatics, 175*, 27–37. https://doi.org/10.1016/j.pragma.2021.01.011

Baron, N. S. (2023a). *Medium matters for reading: What we know about learning with print and digital screens: A report for K-12 educators* [Online resource]. American University. https://doi.org/10.57912/23857212.v1

Baron, N. S. (2023b, January 19). How ChatGPT robs students of motivation to write and think for themselves. *The Conversation*.

Bean, J. C. (2011). *Engaging ideas: The professor's guide to integrating writing, critical thinking, and active learning in the classroom* (2nd ed.). Jossey-Bass.

Bean, J. C., & Melzer, D. (2021). *Engaging ideas: The professor's guide to integrating writing, critical thinking, and active learning in the classroom* (3rd ed.). Jossey-Bass.

Behizadeh, N. (2014). Adolescent perspectives on authentic writing instruction. *Journal of Language & Literacy Education, 10*(1), 27–44.

Behizadeh, N. (2019). Aiming for authenticity: Successes and struggles of an attempt to increase authenticity in writing. *Journal of Adolescent & Adult Literacy, 62*(4), 411–419. https://doi.org/10.1002/jaal.911

Berninger, V., Abbott, R., Jones, J., Wolf, B., Gould, L., Anderson-Youngstrom, M., Shimada, S., & Apel, K. (2006). Early development of language by hand: Composing-, reading-, listening-, and speaking- connections, three letter writing modes, and fast mapping in spelling. *Developmental Neuropsychology, 29*, 61–92.

Buber, M. (1947). *Between man and man* (R. G Smith, Trans.). Routledge & Kegan Paul.

Canada. (2023a). Voluntary Code of Conduct on the Responsible Development and Management of Advanced Generative AI Systems, Innovation, Science and Economic Development Canada, September 2023. https://ised-isde.canada.ca/site/ised/en/voluntary-code-conduct-responsible-development-and-management-advanced-generative-ai-systems. Accessed 8 Dec 2023.

Canada. (2023b). Frequently asked questions for the Voluntary Code of Conduct on Advanced Generative AI Systems, Innovation, Science and Economic Development Canada, September 2023. https://ised-isde.canada.ca/site/ised/en/frequently-asked-questions-voluntary-code-conduct-advanced-generative-ai-systems. Accessed 8 Dec 2023.

Canada. (2023c). *The Artificial Intelligence and Data Act (AIDA) – Companion document*. (2023, March 13). Innovation, Science and Economic Development Canada. https://ised-isde.canada.ca/site/innovation-better-canada/en/artificial-intelligence-and-data-act-aida-companion-document

Çavdar, G., & Doe, S. (2012). Learning through writing: Teaching critical thinking skills in writing assignments. *PS, Political Science & Politics, 45*(2), 298–306. https://doi.org/10.1017/S1049096511002137

Chan, C. K. Y. (2023). A comprehensive AI policy education framework for university teaching and learning. *International Journal of Educational Technology in Higher Education, 20*(1), 38. https://doi.org/10.1186/s41239-023-00408-3

Digital Life Institute (DLI). https://www.digitallife.org/

Eaton, S. E. (2023, February 25). 6 tenets of postplagiarism: Writing in the age of artificial intelligence. *Learning, Teaching and Leadership*.
Ellerton, P., Kuss, A., Vaughan, K., Leibovitch, Y., & Brown, D. (2022). *Teaching critical thinking and why it matters: A transdisciplinary pedagogy for teaching critical thinking*. https://doi.org/10.35542/osf.io/wq5g4
Fraser, D., & Spiller, D. (1996). A pedagogy for improving writing and thinking. *Waikato Journal of Education, 2*(1). https://doi.org/10.15663/wje.v2i1.511
Freire, P. (1970). *Pedagogy of the oppressed*. Seabury.
G'Sell, E. (2024, May 30). Opinion: Algorithms and the problem of intellectual passivity. *The Chronicle of Higher Education*. https://www.chronicle.com/article/algorithms-and-the-problem-of-intellectual-passivity
Garg, M., & Goel, A. (2022). A systematic literature review on online assessment security: Current challenges and integrity strategies. *Computers & Security, 113*, 102544. https://doi.org/10.1016/j.cose.2021.102544
Hargreaves, H., Robin, S., & Caldwell, E. (2022). Student perceptions of reading digital texts for university study. *Journal of Learning Development in Higher Education, 24*, 24. https://doi.org/10.47408/jldhe.vi24.817
Harrison, B. T. (1983). *Learning through writing: Stages of growth in English*. NFER-Nelson.
Higher Education Strategy Associates. (2023). AI Observatory. https://higheredstrategy.com/ai-observatory-home/ai-observatory-policies-and-guidelines/. Accessed 8 Dec 2023.
hooks, b. (2010). *Teaching critical thinking: Practical wisdom*. Routledge. https://books-scholarsportal-info.uproxy.library.dc-uoit.ca/en/read?id=/ebooks/ebooks2/taylorandfrancis/2013-04-15/3/9780203869192#page=198
Hughes, J., & King, A. E. (2010). Dual pathways to expression and understanding: Canadian coming-of-age graphic novels. *Children's Literature in Education, 41*(1), 64–84. https://doi.org/10.1007/s10583-009-9098-8
Hughes, J. M., King, A., Perkins, P., & Fuke, V. (2011). Adolescents and 'autographics': Reading and writing coming-of-age graphic novels. *Journal of Adolescent & Adult Literacy, 54*(8), 601–612.
James, K. H., & Engelhardt, L. (2012). The effects of handwriting experience on functional brain development in pre-literate children. *Trends in Neuroscience and Education, 1*(1), 32–42. https://doi.org/10.1016/j.tine.2012.08.001
Jamieson, S., & Howard, R. M. (2019). Rethinking the relationship between plagiarism and academic integrity. *Revue Internationale Des Technologies En Pédagogie Universitaire, 16*(2), 69–85. https://doi.org/10.18162/ritpu-2019-v16n2-07
Johnson, I. R., & Pietri, E. S. (2023). Ally endorsement: Exploring allyship cues to promote perceptions of allyship and positive STEM beliefs among White female students. *Group Processes & Intergroup Relations, 26*(3), 738–761. https://doi.org/10.1177/13684302221080467
Kelly, S. M., Est, P., & January, T. (2023, January 26). ChatGPT passes exams from law and business schools. *CNN*. https://www.cnn.com/2023/01/26/tech/chatgpt-passes-exams?fbclid=IwAR12aDz2c-2SVKfyS2fozL_VRk41AGavdGKsoP6qtNyCHmpVtLfDGpu4WXg
Khanna, N. (2020). *The visceral logics of decolonization*. Duke University Press.
Kuh, G. D. (2003). What we're learning about student engagement from NSSE. *Change, 35*(2), 24. https://doi.org/10.1080/00091380309604090
Lewis, P. R., & Marsh, S. (2022). What is it like to trust a rock? A functionalist perspective on trust and trustworthiness in artificial intelligence. *Cognitive Systems Research, 72*, 33–49. https://doi.org/10.1016/j.cogsys.2021.11.001
Lewis, P. R., & Sarkadi, Ş. (2024). Reflective artificial intelligence. *Minds and Machines, 34*(2), 14. https://doi.org/10.1007/s11023-024-09664-2
Lewis, P. R., Marsh, S., & Pitt, J. (2021, June 10). AI vs "AI": Synthetic minds or speech acts. *IEEE Technology and Society*. http://technologyandsociety.org/ai-vs-ai-synthetic-minds-or-speech-acts/

Lieberman, L. (2024, June 7). Opinion: AI and the death of student writing. *The Chronicle of Higher Education*. https://www.chronicle.com/article/ai-and-the-death-of-student-writing
Liu, W., Huang, H., Saleem, A., & Zhao, Z. (2022). The effects of university students' fragmented reading on cognitive development in the new media age: Evidence from Chinese higher education. *PeerJ, 10*, e13861. https://doi.org/10.7717/peerj.13861
McDaniel, D. S. (2024). Toward culturally digitized pedagogy: Informing theory, research, and practice. *Reading Research Quarterly*, 1–18. https://doi.org/10.1002/rrq.534
McGill, J. (2023). How many AI tools are there? Content at Scale. Blog. https://contentatscale.ai/blog/how-many-ai-tools-are-there/
McKinsey & Company. (2024, April 2). What is generative AI? Blog. https://www.mckinsey.com/featured-insights/mckinsey-explainers/what-is-generative-ai
McKnight, L., & Shipp, C. (2024). "Just a tool"? Troubling language and power in generative AI writing. *English Teaching: Practice & Critique, 23*(1), 23–35. https://doi.org/10.1108/ETPC-08-2023-0092
Miedema, T. (2023, October 17). The integrity report: A Dr. T Publication, Faculty of Business and Information Technology, Ontario Tech University. Vol. 2, Issue 1.
Mindzak, M., & Eaton, S. E. (2021). Artificial intelligence is getting better at writing, and universities should worry about plagiarism. *The Conversation*.
Mohamed, S., Png, M.-T., & Isaac, W. (2020). Decolonial AI: Decolonial theory as sociotechnical foresight in artificial intelligence. *Philosophy & Technology, 33*(4), 659–685. https://doi.org/10.1007/s13347-020-00405-8
Moya, B., Eaton, S., Pethrick, H., Hayden, A., Brennan, R., Wiens, J., & McDermott, B. (2024). Academic integrity and artificial intelligence in higher education (HE) contexts: A rapid scoping review. *Canadian Perspectives on Academic Integrity, 7*(3), 3. https://doi.org/10.55016/ojs/cpai.v7i3.78123
Mueller, P. A., & Oppenheimer, D. M. (2014). The pen is mightier than the keyboard: Advantages of longhand over laptop note taking. *Psychological Science, 25*(6), 1159–1168.
Nafukho, F. M. (2023, March 25). AI and the importance of meaningful humane learning. *University World News*.
Nascimento, F., & Chown, E. (2023). *Meaningful technologies: How digital metaphors change the way we think and live*. Lever Press. https://muse.jhu.edu/pub/314/oa_monograph/book/110606
Noble, S. U. (2018). *Algorithms of oppression: How search engines reinforce racism*. New York University Press.
Omizo, R., & Hart-Davidson, B. (2024). Is genre enough? A theory of genre signaling as generative AI rhetoric. *Rhetoric Society Quarterly, 54*(3), 272–285. https://doi.org/10.1080/02773945.2024.2343615
Pedersen, I. (2023). The rise of generative AI and enculturating AI writing in postsecondary education. *Frontiers in Artificial Intelligence, 6*. https://doi.org/10.3389/frai.2023.1259407
Pedersen, I. (2024). Generative AI adoption in postsecondary education, AI hype, and ChatGPT's launch. *The Open/Technology in Education, Society, and Scholarship Association Journal, 4*(1), 1–19. https://doi.org/10.18357/otessaj.2024.4.1.59
Perez-Felkner, L., McDonald, S.-K., Schneider, B., & Grogan, E. (2012). Female and male adolescents' subjective orientations to mathematics and the influence of those orientations on postsecondary majors. *Developmental Psychology, 48*(6), 1658–1673. https://doi.org/10.1037/a0027020
Peters, M., Boies, T., & Morin, S. (2019). Teaching academic integrity in Quebec universities: Roles professors adopt. *Frontiers in Education, 4*, 1–13.
Rogerson, A. M., & McCarthy, G. (2017). Using Internet based paraphrasing tools: Original work, patchwriting or facilitated plagiarism? *International Journal for Educational Integrity, 13*(1), 1. https://doi.org/10.1007/s40979-016-0013-y
Rutter, M. P., & Mintz, S. (2023, January 16). ChatGPT: Threat or menace? *Higher Ed Gamma, Inside Higher Ed*. https://www.insidehighered.com/blogs/higher-ed-gamma/chatgpt-threat-or-menace

Salajan, F. D., Barnes, T. L., & Becker, A. (2024). Comparative regionalism cases of artificial intelligence governance in education: The Caribbean Community and the European Union. *The Round Table, 113*(1), 71–85. https://doi.org/10.1080/00358533.2024.2307793

Schnitzler, C., Vee, A., & Laquintano, T. (Eds.). (2024). TextGenEd: Continuing experiments. *The WAC Repository, 2*(1), 1–150. https://doi.org/10.37514/TWR-J.2024.2.1.01

Sharples, M. (2023). Towards social generative AI for education: Theory, practices and ethics. *Learning: Research and Practice, 9*(2), 2.

Singer, L. M., & Alexander, P. A. (2017). Reading across mediums: Effects of reading digital and print texts on comprehension and calibration. *The Journal of Experimental Education, 85*(1), 155–172. https://doi.org/10.1080/00220973.2016.1143794

Strzelecki, A. (2023). Students' acceptance of ChatGPT in higher education: An extended unified theory of acceptance and use of technology. *Innovative Higher Education*. https://doi.org/10.1007/s10755-023-09686-1

Université de Montréal. (2018). Montréal Declaration for a Responsible Development of Artificial Intelligence (Montréal Declaration). https://montrealdeclaration-responsibleai.com/

Vasset, F. P., Dahl, B. M., Thunem, G., & Frilund, M. (2024). Exploring scientific writing as part of a learning model for interprofessional higher health education – A qualitative study. *Social Sciences & Humanities Open, 9*, 100759. https://doi.org/10.1016/j.ssaho.2023.100759

Vee, A. (2023). Large language models write answers. *Composition Studies, 51*(1), 176–182.

Watson, G. P. L., & Eaton, S. E. (2023, February 17). AI tools don't have to be the enemy of teaching and learning. *University Affairs*.

Wei, F. (2021). Cultivation of students' critical thinking in the era of artificial intelligence-based on the perspective of training mode and education subjects. In V. Sugumaran, Z. Xu, & H. Zhou (Eds.), *Application of intelligent systems in multi-modal information analytics* (Vol. 1385, pp. 786–790). Springer International Publishing AG. https://books-scholarsportal-info.uproxy.library.dc-uoit.ca/en/read?id=/ebooks/ebooks6/springer6/2021-06-11/4/9783030748142

Wertheimer, T. (2024, July 26). South Korea wrongly introduced as North Korea at Olympics. *BBC.Com*. https://www.bbc.com/news/articles/cmj2r1403jpo

Wu, R., & Yu, Z. (2024). Do AI chatbots improve students learning outcomes? Evidence from a meta-analysis. *British Journal of Educational Technology, 55*(1), 10–33. https://doi.org/10.1111/bjet.13334

Zaphir, L., & Hansen, D. (2024, April 19). The trouble with Bloom's taxonomy in an age of AI. *THE Campus Learn, Share, Connect*. https://www.timeshighereducation.com/campus/trouble-blooms-taxonomy-age-ai

Zembylas, M. (2021). Refusal as affective and pedagogical practice in higher education decolonization: A modest proposal. *Teaching in Higher Education, 26*(7/8), 953–968. https://doi.org/10.1080/13562517.2021.1900816

Zinsser, W. (1988). *Writing to learn*. HarperCollins.

Zull, J. E. (2002). *The art of changing the brain: Enriching the practice of teaching by exploring the biology of learning*. Stylus.

Chapter 7
Surveillance and Guidance: An Academic Integrity Crossway for Humans Using Generative AI

Gabriel Dumouchel and Martine Peters

Abstract While plagiarism cases are on the rise since the advent of generative artificial intelligence (GenAI), higher education institutions must determine how they will deal with this new reality. This chapter explores how to prevent plagiarism when GenAI is used in academic writing. The Academic Integrity Crossway model puts prevention at the heart of a combined approach of surveillance and guidance where students' written assignments are channeled through prevention. Surveillance—in the form of plagiarism detection or sanctions—is needed to deter academic misconduct while guidance—in the form of pedagogical methods and interventions such as restorative practices—is needed to foster academic integrity. Our model furthers this coalition by adding AI to contribute to every part of the prevention process. On the one hand, AI can assist institutions for surveillance purposes with detection and management tools, uncovering plagiarism cases and improving how they are handled. On the other hand, AI can provide guidance by serving as academic integrity tutors to students while they are writing as well as their academic integrity mentors after they've been caught plagiarizing. Our model shows that plagiarism prevention and academic integrity promotion are complementary as positive actions made by humans and GenAI are best suited to achieve a flourishing culture of academic integrity.

Keywords Academic integrity · Generative artificial intelligence · Guidance · Higher education · Surveillance

G. Dumouchel (✉)
Université du Québec à Chicoutimi, Chicoutimi, QC, Canada
e-mail: gabriel_dumouchel@uqac.ca

M. Peters
Université du Québec en Outaouais, Gatineau, QC, Canada
e-mail: martine.peters@uqo.ca

A. E. King (ed.), *Artificial Intelligence, Pedagogy and Academic Integrity*, Advances in Artificial Intelligence in Education 1,
https://doi.org/10.1007/978-3-031-92534-4_7

Introduction

Written papers and academic essays are a mainstay of higher education assessment (Graves et al., 2010), as they develop students' argumentation skills and are easy to set up for assessing the depth of their understanding (Sharples, 2022). Unfortunately, these type of assignments are challenging for some students who lack writing skills, which can lead them to plagiarize (Hafsa, 2019). Most institutions have plagiarism policies with committees in place who sanction students found guilty of academic misconduct (Yankova, 2024). However, with generative artificial intelligence (GenAI) use on the rise in higher education, these policies must be rewritten to include plagiarism with the help of GenAI and how to deal with it (Dwivedi et al., 2023). This chapter explores how to prevent plagiarism when GenAI is used in academic writing by proposing a model combining a surveillance and guidance approach where students' written assignments are channeled through prevention. Our model furthers this coalition by adding AI to contribute to every part of the prevention process.

Academic Writing and Plagiarism

According to Peters (2015), academic writing is the result of a recursive writing process where a written product emerges from a series of iterative actions of planning, writing, and revision/correction. This process combines the implementation of three complementary types of skills. First, informational skills allow a text to be well documented by identifying the subject studied to better search, find and sort the information that meets the targeted need. Second, writing skills allow the appropriation of said documentation (i.e., note-taking, summary, interpretation), its integration into the creation of a new text as well as its polishing through rewriting, correction, revision and formatting. Third, referencing skills solidify the text by acknowledging sources used through quotations, paraphrases and bibliography. These three skills can be performed individually or collaboratively in the context of academic writing (Association of College and Research Libraries, 2015).

However, at any point during the process of writing an assignment, students may plagiarize, intentionally or not (Camara et al., 2017). For example, students using their informational skills may forget to note the source of the information they found and submit their assignments without properly referencing it. Improperly referencing information is a form of plagiarism (Lindahl & Grace, 2018). Another very common type of plagiarism is copying and pasting directly from the text of another author. Passing off information as one's own is another form of plagiarism (Gipp, 2014).

Traditional Plagiarism Versus a New Form of Plagiarism

Traditionally, plagiarism has been defined as taking someone else's words or ideas and passing them off as one's own (Dougherty, 2018; Fishman, 2009). The arrival of ChatGPT in the public sphere changes the definition of plagiarism. Peters (2023) has published a new definition which includes GenAI: plagiarism is presenting the words or ideas of another person, *or those generated by an AI*, without reference to the source from which the information comes, with the aim of deriving benefit from them in an evaluation context (paragraph 10).

GenAI, a Plagiarism or Learning Tool

GenAI writing tools present several challenges to academic integrity. Perkins and Roe (2023) note that it is now very fast and easy to cheat by generating parts of or whole assignments without proper attribution, a new reality that can weaken the educational value of academic tasks (Strzelecki, 2023) as well as academic institutions' credibility (Bretag, 2013). GenAI tools can also fabricate content like fake sources and nonsensical information (Crompton & Burke, 2024), which represents a major problem for academic integrity if it is used by students in their writing assignments. Furthermore, as GenAI tools become mainstream and reliable, Nikolic et al. (2024) warn that students could be more tempted to cheat, be it intentionally or not, as such technology becomes intertwined in their regular lives.

On the other hand, according to Polyportis and Pahos (2024), students perceive the relative easiness of GenAI tools and their potential as a way to enhance their learning outcomes and academic performance. Li et al. (2024) showed that large language model (LLM) chatbots like ChatGPT can provide personalized feedback, enhance writing motivation and significantly improve students' writing performance in terms of content quality and language expression. In their systematic review, Crompton and Burke (2024) noted that ChatGPT could offer different kinds of writing support for students such as identifying and correcting grammatical errors, brainstorming writing topics, producing a first draft and assisting in alternative phrasing. The scientific literature shows that GenAI can both serve as a learning and a plagiarizing tool. It all depends on the way students use it in higher education settings. Prevention plays a key role to favor ethical behavior over academic misconduct but to do so efficiently in the fast-changing landscape of LLMs is still up to debate.

Preventing Plagiarism and Promoting Academic Integrity in the Age of GenAI

The arrival of GenAI has created a tsunami in education (Tan et al., 2024), particularly an increase in plagiarism cases in higher education institutions (Wiley, 2024). Institutions need to react to this GenAI plagiarism phenomenon. In the past, most institutions adopted a sanctioning approach preventing plagiarism (Macdonald & Carroll, 2006; Park, 2003). However, research has found that this approach is not sufficient to deter plagiarism (Awad et al., 2013; Freiburger et al., 2017; Ogilvie & Steward, 2010).

One solution resides in **reframing academic integrity** for all involved, from the institution to its staff, faculty and students. Luo (2024a) proposes reconsidering the originality of students' work in the era of GenAI. Instead of stigmatizing its use in their assignments, higher education policies should reframe originality from a collaborative perspective set on a continuous spectrum, where an original piece of work is the end product of a fluctuating collaborative effort between students and GenAI (Luo, 2024a).

Ellis and Murdoch (2024) recommend that higher education institutions use responsive regulation with cheating, balancing when to punish and when to persuade students. Their educational integrity enforcement pyramid combines a spectrum of students' attitudes toward cheating (from champions of integrity to criminals) and a responsive institutional strategy that ranges from support to compulsion. Their idea is that institutions must be both ferocious and forgiving to challenge cheating.

Eaton (2023) goes beyond GenAI's impact on academic integrity with what she coined the postplagiarism era; that is, neuroeducational-technology will eventually render detection obsolete as technology will be invisible and fully integrated in everyday life. The technology is not yet at that stage so for the moment, instructors and institutions need to consider how a culture of academic integrity can be built and championed in a pragmatic and positive way. To achieve this, we agree with the suggestion of Ellis and Murdoch (2024) that the debate needs to move beyond adversarial (detection and sanctions) versus cooperative (pedagogy and interventions) approaches and accept that institutions need to do both in a balanced way in order to address academic misconduct successfully.

We thus propose the following Academic Integrity Crossway model (see Fig. 7.1) where prevention is at the heart of a combined surveillance and guidance approach where students' written assignments are channeled through prevention, be it via detection, sanctions, pedagogy and interventions. Surveillance is needed to prohibit academic misconduct while guidance is needed to foster academic integrity and so both are needed to prevent plagiarism. Our model furthers this coalition with the addition of GenAI to contribute to the prevention process.

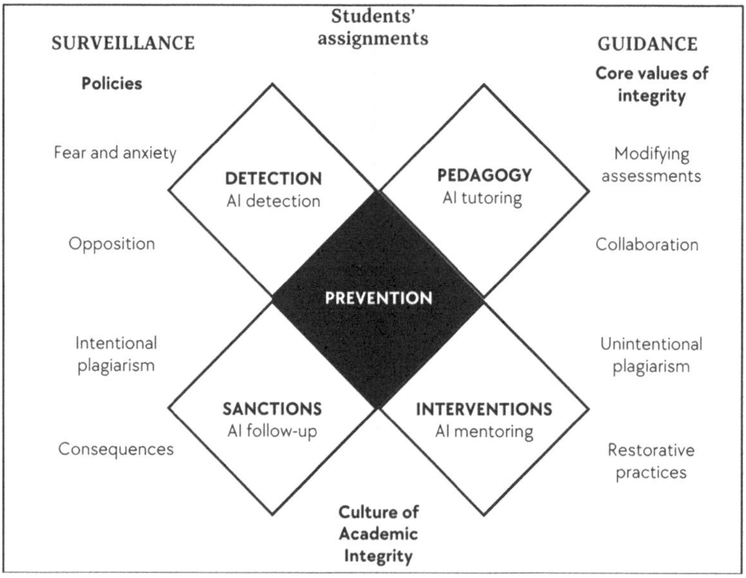

Fig. 7.1 The Academic Integrity Crossway

Prevention at the Heart of a Culture of Academic Integrity

Every institution wants to have a culture of academic integrity within their walls. Yet it is not an easy thing to establish especially since our society has seen a general loosening of morals in the last few decades (Barnhardt, 2016); for example, there tends to be little respect for authors' rights and intellectual property. Preventing plagiarism in education is the responsibility of everyone in the institution (Walker & White, 2014). One of the most efficient ways to preserve academic integrity is by demanding original and authentic assignments which will develop critical thinking and communication skills in students (Sokhanvar et al., 2021) and ensure that assessments are fair and equitable for all (Suskie, 2018). This in turn will maintain the reputation and credibility of higher education institutions (Thomson et al., 2023). It is the responsibility of higher education institutions to promote intellectual discipline and develop ethical skills in their students (Scanlan, 2006) to foster honest, fair and productive citizens. That is why plagiarism prevention is at the heart of our combined approach.

A Combined Approach to Prevention: Surveillance and Guidance

Our model offers two paths that can be taken when a written assignment is submitted to a teacher. On the one hand, prevention can be done with a focus on **surveillance**, as institutional policies are applied when academic misconduct is detected and sanctioned, thus fostering fear, anxiety and a culture of distrust between students and teachers.

On the other hand, prevention can focus on **guidance**, where core values of academic integrity are taught and interventions are made when there is academic misconduct, thus creating a culture of academic integrity for students and teachers. In the following paragraphs, we describe and argue the components and impacts of both paths, while also considering the complementary roles of humans and GenAI to prevent academic misconduct.

Surveillance: The Traditional but Still Needed Part of the Combined Approach

For prevention to be effective, an institution must develop, implement and disseminate academic integrity **policies** to the entire community, from students to teachers and support staff such as librarians. The surveillance approach is based on these policies. However, with the arrival of GenAI, some researchers argue that institutional policies need to be reviewed because they lack clarity regarding the authorized use of GenAI and are not well understood and known by students (Gorichanaz, 2023; Polyportis & Pahos, 2024). In this regard, Yusuf et al. (2024) emphasize that appropriate policies should be put in place to balance the potential benefits of GenAI while maintaining academic integrity. Perkins and Roe (2023) argue that policies must be nuanced and multifaceted to adapt to the context and priorities of each institution. Thus, some of them will favor a more restrictive approach—prohibiting the use of GenAI—while others will be more permissive by focusing on an ethical and transparent use of these tools by students.

However, the mere existence of such policies means that students are not necessarily trusted to act with integrity during their education. Many of them may perceive mistrust from their institution and teachers, which can create a climate of **fear and anxiety** that is contrary to a healthy learning environment (Luo, 2024b). On one hand, students will not trust teachers monitoring and threatening them with academic sanctions; and on the other hand, teachers will think students are trying to dupe them by cheating in their assignments, creating a climate of **opposition**. This can cause negative emotions which in turn can lead to **intentional plagiarism** (Tindall et al., 2021). The objective of a combined approach to plagiarism prevention is to eliminate this adversarial climate, reassure students by involving all members of the community in the development of academic integrity policies and to promote inclusion and trust between the various parties concerned.

Detection and Sanction

Every institution tries to prevent plagiarism with **detection** methods that are done manually or by automation. Instructors can always bear the brunt of this job by trying to detect cases of plagiarism in submitted assignments, but manual detection represents a tedious and time-consuming task (Bretag & Mahmud, 2009).

Plagiarism detection software like Turnitin or Compilatio greatly accelerate the verification process (Weber-Wulff, 2015), but in the age of GenAI, these tools have been found to be insufficiently reliable (Farrelly & Baker, 2023). They can make mistakes by producing false positives and false negatives, which can lead to wrongly accusing or exonerating students. Furthermore, Farrelly and Baker (2023) note that the creation and use of detection tools amounts to defending academic integrity through techno-solutionism rather than education. Cotton et al. (2024) compare it to an arms race where no one is certain who will win. But with pragmatism in mind, we can simply state that if there is a market for plagiarism, be it with contract cheating services (Rigby et al., 2015) or GenAI tools, there is also one for countering it (McMurtrie, 2019). Moreover, even when plagiarism is detected by hand or machine in a student's assignment, it needs to be proven that it was done intentionally. This process can be long and hard, which in turn makes many teachers refuse to confront students directly or report a case to the institution (Sopcak & Hood, 2022). It is a game of cat and mouse where both fear each other, while running in a permanently altering maze of advancing technologies. Because surveillance is tied to negative human emotions like fear and anxiety coupled with a top-down relationship, detection puts students and teachers in opposition. But manual and automated detection must still be done, as it is part of plagiarism prevention where students learn about the disciplinary process; many will not plagiarize since they know they can be caught, and sanctions applied (Zhang et al., 2024).

When intentional plagiarism is confirmed by gathering sufficient evidence, instructors and institutions can use approved **sanctions** to impose consequences on the offending student (Tarisayi, 2023). This approach may punish students by failing an assignment, a course, or even—in serious or repeated cases—by a temporary or definitive expulsion from a training program. While humans ultimately decide the fate of a sanctioned student, GenAI could serve as an institutional follow-up management tool, gathering information on how every plagiarism case is processed and generating improvement proposals on how to deal with future cases.

In our combined model, plagiarism detection and sanctions are as important as pedagogy and interventions for building up a culture of academic integrity. Students need to understand that they play an important role in building this culture as collaborators, not opponents. To reduce their adversarial stance (opposition) and negative perception of their situation (fear and anxiety), students should act as plagiarism prevention collaborators with their institution and teachers, helping to choose what should be done to detect plagiarism and what kind of sanctions should be brought to offenders. A sense of community building with common goals of integrity can emerge from such a mindset.

Guidance: A Path Less Traveled but Equally Important

If the surveillance approach is based on institutional policies, the guidance approach is based on the core values of academic integrity. These core values, as defined by the International Center for Academic Integrity (ICAI, 2021), are the foundational elements needed to build a culture of academic integrity—honesty, trust, fairness, respect, responsibility, and courage. For example, fairness will only work if teachers apply rules and policies consistently while students take responsibility for their own actions. Both actors are dancing on the same stage, and the quality of their choreography depends on the complementary nature of their actions. These core values are articulated in the policies and pedagogies put in place by an academic institution. The ICAI emphasizes that policies on academic integrity must be clear to be well understood and applied, in addition to being promoted in a positive way. All members of the community need to be trained on these policies so that expectations become part of their shared culture. The ICAI also suggests that the pedagogy employed by teachers allows students to learn from their mistakes in terms of academic integrity. This can be linked to restorative practices (RP) where students are given the opportunity to acknowledge their wrongdoing, take responsibility, and participate in community building (Sopcak & Hood, 2022).

However, Luo (2024b) says that scholarly institutions tend to address academic integrity in a negative way, putting the focus on identifying and prohibiting academic misconduct, which in turn means focusing on detecting and sanctioning bad behaviors. Surveillance remains necessary to deter plagiarism and enforce consequences to foster academic integrity and assessment security at the institutional level (Ellis & Murdoch, 2024). Yet, according to Luo (2024b), when GenAI policies are presented to students in a surveillance tone, it can undermine trust between them and their teachers who will be perceived as gatekeepers of academic misconduct rather than educators. Luo (2024b) proposes that scholarly institutions should adopt a more supportive and inclusive approach with GenAI tools, where policies and measures guide students toward an ethical use of this technology for their learning experience.

Pedagogy and Interventions

When prevention is focused on guidance, **pedagogy** can ensure that the institution and its teachers make judicious choices in terms of teaching and learning assessment to promote academic integrity. Each teacher must remind students of the rules related to the respect of academic integrity with or without GenAI. They can emphasize why this is important for their learning process and for the construction of their professional identity. Students must be aware of the impact of a lack of integrity in an academic context, but also in the profession concerned by their studies (Nikolic et al., 2024). Examples of cases reported in traditional media can illustrate the real consequences of not demonstrating integrity in professional settings. A law teacher could thus present situations where lawyers have been sanctioned for misusing GenAI (Merken, 2023; Proctor, 2024). Such examples can foster awareness for students that they will always be responsible for their actions. In short, professional integrity must be seen as the natural extension of academic integrity to discourage plagiarism.

Furthermore, **pedagogical methods** promoting the active participation of students, such as workshops and debates, are better suited at curbing the use of plagiarism than more traditional methods such as lectures, where students are more passive (Michel-Villarreal et al., 2023). But as the popularity of GenAI tools grows exponentially with students, it is even more pressing for scholarly institutions to provide support for instructors to reassess and redesign their courses and **assessment methods**, as many are predicting the end of the essay as a traditional form of academic assessment (Cotton et al., 2024; Stokel-Walker, 2022; Yeadon et al., 2023). Assessment of students' assignments written with or by GenAI must now focus on the writing process more than the submitted work (see Chap. 6 by King and Garramone). The goal is to get students to demonstrate that they truly achieved the learning objectives put forth by their institution. Therefore, to prevent plagiarism, teachers need to use various assessment methods focusing on the writing process, combining traces, sources, reflections and in-person verifications. They should also gain trust by perceiving students as collaborators rather than potential offenders to build a culture of academic integrity together. Luo (2024b) suggests that teachers could ask students how they think GenAI should be accounted for in assessment tasks.

Most GenAI writing tools are not yet calibrated to **tutor** students to follow academic integrity principles and rules. In an ideal world, they could help students to write with integrity in real time, for example, with reminders to give their sources, or to properly quote authors. For the moment, breaches of integrity are mainly uncovered by detection tools after a writing task has been completed by the student and submitted for academic assessment. While GenAI tools are said to have tremendous potential for tutoring, especially for problem solving (Nikolic et al., 2024), we can safely say that it also has the potential to serve as academic integrity **tutors** for students during their writing process. Indeed, by integrating institutional policies directly into GenAI tools, students could receive policy reminders or guidance while they are writing. Such tutoring could be displayed in the query interface of a GenAI tool in relation to the quality of the information generated,[1] as well as in the generated texts themselves. It is essential that GenAI tools like ChatGPT and

[1] On July 25, 2024, the following message appeared under the ChatGPT query bar: "ChatGPT can make mistakes. Consider checking important information."

Copilot promote academic integrity to users in general and to students more specifically. If modern cars can remind drivers to keep their hands on the wheel, GenAI tools can remind students to keep academic integrity in mind during their writing process.

As pedagogy can only do so much to prevent plagiarism, our guidance path can offer offending students a second chance with an **intervention**. According to Sopcak and Hood (2022), while a punitive approach may still be needed in some cases of academic misconduct, research has shown that a quasi-legal approach can lead to a widespread unwillingness for teachers to report it to their institution. Having an adversarial approach also causes shame and stigmatization of wrongdoers, thus contributing to a culture of distrust between teachers and students. Rather than only inducing fear and rule-compliance among at-fault students, being caught plagiarizing can be used as positive leverage for academic integrity. Teachable moments and restorative practices (RP) can transform such moments into experiential learning opportunities that promote student accountability.

When plagiarism arises from lack of knowledge on the part of students, teachers can take a teachable moment to explain to them which mistakes have been made that have led to unintentional plagiarism (Fisk, 2024). These teachable moments are unique opportunities to make learning about plagiarism prevention more relevant to students. These teachable moments should be included in the institutional plagiarism policies.

Sopcak (2020) summarizes RP as a relatively structured process in which a trained facilitator leads the responsible party (i.e., the at-fault student) and the harmed parties (i.e., the instructor and a student association representative when appropriate) through a series of questions that compel students to consider the material, emotional and community effects of their actions and assist them in taking responsibility. Sopcak and Hood (2022) add that while academic integrity policies can be advocated and rewarded when followed, academic integrity values are better internalized when it stems from moral reflection. Sopcak and Hood argue that as part of the moral and civic education of students, an institution should encourage respect of rules and regulations and fostering remorse, as well as helping them right the wrongs they have done to their instructors, themselves and their community. Moreover, Jobidon (2024) emphasizes that while instructors are limited in their assessment of their students' intention to plagiarize, they are still in the best position to intervene. However, this type of intervention should be limited to "lighter" cases where students did not intend to plagiarize or have been negligent or forgetful in the wake of academic integrity (Fisk, 2024). Using teachable moments and RP can raise awareness for both students and teachers that academic integrity can serve to build healthy relationships and communities based on trust and responsibilities.

Köbis and Mehner (2021) have been pondering about the use of AI-supported mentoring in higher education. If critical issues such as data privacy and protection are correctly addressed, we argue that GenAI tools could also eventually act as integrity mentors for students caught plagiarizing. While teachers can discuss in-person or online with their students about their academic misconduct, a GenAI mentoring tool could offer additional support in a 24/7 capacity, ensuring a follow-up tailored for their personal needs.

While detection and sanctions need students to act as plagiarism prevention collaborators to build a culture of academic integrity, pedagogy and interventions make them active promoters of said culture. Therefore, assessments are not only modified to respond to GenAI's threat on academic integrity; they are modified with students' input to legitimize them and improve their learning value and their positive impact on academic integrity. Restorative practices can make offenders champions of redemption, helping them to bring awareness to others and consequently bring down the number of unintentional plagiarism cases.

Our model shows that plagiarism prevention and academic integrity promotion are intertwined by a crossway of surveillance and guidance paths. Both ways are complementary as positive actions made by humans and GenAI are best suited to achieve a flourishing culture of academic integrity.

Conclusion

The fundamental goal of higher education institutions is to impart knowledge, foster personal and intellectual development, and prepare students for lifelong learning. This comprehensive role played by higher education institutions means that they must also promote ethical behavior and social responsibility to their students. With GenAI use on the rise in higher education institutions, it is more important than ever to promote academic integrity. Students are there to develop skills and acquire knowledge needed to become productive members of their profession. GenAI can be used ethically in many ways to help in written assignments: for example, to search for information or properly reference sources. However, some students might choose the easy way and use GenAI to plagiarize. It is our duty to ensure that all assignments that are graded are genuine and represent the true effort of students. Encouraging them to produce creative and original assignments is vital for their learning but also to making them become productive and honest members of our society.

In order to do so, higher education institutions ought to prevent plagiarism by combining a surveillance and a guidance approach. We argue that prevention comes down to balancing the yin (deterring academic misconduct) and the yang (encouraging academic integrity). Higher education institutions cannot rely solely on detection and sanctions, nor can they only rely on pedagogy and interventions. Sopcak and Hood (2022) illustrate this duality by arguing that restorative practices should be the default response to academic misconduct while a punitive approach should remain a secondary option when conditions for the latter approach aren't met. Building a strong and perennial culture of academic integrity needs to rely on pragmatism and flexibility.

Furthermore, while GenAI can be used as a tool to plagiarize in higher education, it can also help students in their educational endeavors. The same can be said for preventing plagiarism, as GenAI can assist a surveillance approach by automatically detecting occurrences of plagiarism (intentional and non-intentional) and offer follow-ups on sanctions, as well as supporting a guidance approach by offering tutoring to learn academic integrity and mentoring to amend academic misconduct. GenAI may not yet be able to do so in a sustainable and reliable way, but like a hammer can build or break things, it represents a tool that higher education will have to develop, master and use wisely to create ensure a thriving culture of academic integrity.

References

Association of College and Research Libraries. (2015). Framework for information literacy for higher education. https://www.ala.org/acrl/standards/ilframework

Awad, M. K., Zogheib, B., & Alazemi, H. M. K. (2013). A penalty scheme for academic dishonesty. In M. J. W. Lee (Ed.), *Proceedings of 2013 IEEE international conference on teaching, assessment and learning for engineering* (pp. 580–584). https://doi.org/10.1109/TALE.2013.6654504

Barnhardt, B. (2016). The "epidemic" of cheating depends on its definition: A critique of inferring the moral quality of "cheating in any form". *Ethics & Behavior, 26*(4), 330–343. https://doi.org/10.1080/10508422.2015.1026595

Bretag, T. (2013). Challenges in addressing plagiarism in education. *PLoS Medicine, 10*(12), e1001574. https://doi.org/10.1371/journal.pmed.1001574

Bretag, T., & Mahmud, S. (2009). A model for determining student plagiarism: Electronic detection and academic judgement. *Journal of University Teaching and Learning Practice, 6*(1), 49–60. https://doi.org/10.53761/1.6.1.6

Camara, S., Eng-Ziskin, S., Wimberley, L., Dabbour, K. S., & Lee, C. M. (2017). Predicting students' intention to plagiarize: An ethical theoretical framework. *Journal of Academic Ethics, 15*(1), 43–58. https://doi.org/10.1007/s10805-016-9269-3

Cotton, D. R. E., Cotton, P. A., & Shipway, J. R. (2024). Chatting and cheating: Ensuring academic integrity in the era of ChatGPT. *Innovations in Education and Teaching International, 61*(2), 228–239. https://doi.org/10.1080/14703297.2023.2190148

Crompton, H., & Burke, D. (2024). The educational affordances and challenges of ChatGPT: State of the field. *TechTrends, 68*(2), 380–392. https://doi.org/10.1007/s11528-024-00939-0

Dougherty, M. V. (2018). *Correcting the scholarly record for research integrity: In the aftermath of plagiarism.* Springer International. https://doi.org/10.1007/978-3-319-99435-2

Dwivedi, Y. K., Kshetri, N., Hughes, L., Slade, E. L., Jeyaraj, A., Kar, A. K., … Wright, R. (2023). Opinion paper: "So what if ChatGPT wrote it?" multidisciplinary perspectives on opportunities, challenges and implications of generative conversational AI for research, practice and policy. *International Journal of Information Management, 71*, 102642. https://doi.org/10.1016/j.ijinfomgt.2023.102642

Eaton, S. E. (2023). Postplagiarism: Transdisciplinary ethics and integrity in the age of artificial intelligence and neurotechnology. *International Journal for Educational Integrity, 19*(1), 1–10. https://doi.org/10.1007/s40979-023-00144-1

Ellis, C., & Murdoch, K. (2024). The educational integrity enforcement pyramid: A new framework for challenging and responding to student cheating. *Assessment & Evaluation in Higher Education, 49*, 924–934. https://doi.org/10.1080/02602938.2024.2329167

Farrelly, T., & Baker, N. (2023). Generative artificial intelligence: Implications and considerations for higher education practice. *Education Sciences, 13*(11), 1109. https://doi.org/10.3390/educsci13111109

Fishman, T. (2009). "We know it when we see it" is not good enough: Toward a standard definition of plagiarism that transcends theft, fraud, and copyright. In *4th Asia Pacific conference on educational integrity, University of Wollongong NSW Australia*. https://ro.uow.edu.au/apcei/09/papers/37/

Fisk, G. D. (2024). AI or human? Finding and responding to artificial intelligence in student work. *Teaching of Psychology*. https://doi.org/10.1177/00986283241251855

Freiburger, T. L., Romain, D. M., Randol, B. M., & Marcum, C. D. (2017). Cheating behaviors among undergraduate college students: Results from a factorial survey. *Journal of Criminal Justice Education, 28*(2), 222–247. https://doi.org/10.1080/10511253.2016.1203010

Gipp, B. (2014). *Citation-based plagiarism detection*. Springer Fachmedien Wiesbaden. https://link.springer.com/book/10.1007/978-3-658-06394-8

Gorichanaz, T. (2023). Accused: How students respond to allegations of using ChatGPT on assessments. *Learning: Research and Practice, 9*(2), 183–196. https://doi.org/10.1080/23735082.2023.2254787

Graves, R., Hyland, T., & Samuels, B. M. (2010). Undergraduate writing assignments: An analysis of syllabi at one Canadian college. *Written Communication, 27*(3), 293–317. https://doi.org/10.1177/0741088310371635

Hafsa, N.-E. (2019). An overview of theories explaining plagiaristic behavior of students. *Journal of Education and Practice, 10*(25), 102–108. https://doi.org/10.7176/JEP/10-25-11

International Center for Academic Integrity. (2021). *The fundamental values of academic integrity* (3rd ed.). https://academicintegrity.org/resources/fundamental-values

Jobidon, N. (2024). *La sanction du plagiat -- démarquer le rôle de l'enseignant de celui du comité de discipline* [Conference presentation]. Joint International Conference on Ethics and Integrity in Academia, Gatineau, Canada.

Köbis, L., & Mehner, C. (2021). Ethical questions raised by AI-supported mentoring in higher education. *Frontiers in Artificial Intelligence, 4*. https://doi.org/10.3389/frai.2021.624050

Li, H., Wang, Y., Luo, S., & Huang, C. (2024). The influence of GenAI on the effectiveness of argumentative writing in higher education: Evidence from a quasi-experimental study in China. *Journal of Asian Public Policy*. https://doi.org/10.1080/17516234.2024.2363128

Lindahl, J. F., & Grace, D. (2018). Students' and supervisors' knowledge and attitudes regarding plagiarism and referencing. *Research Integrity and Peer Review, 3*, 10. https://doi.org/10.1186/s41073-018-0054-2

Luo, J. (2024a). A critical review of GenAI policies in higher education assessment: A call to reconsider the "originality" of students' work. *Assessment & Evaluation in Higher Education, 49*(5), 651–664. https://doi.org/10.1080/02602938.2024.2309963

Luo, J. (2024b). How does GenAI affect trust in teacher-student relationships? Insights from students' assessment experiences. *Teaching in Higher Education*. https://doi.org/10.1080/13562517.2024.2341005

Macdonald, R., & Carroll, J. (2006). Plagiarism — A complex issue requiring a holistic institutional approach. *Assessment & Evaluation in Higher Education, 31*(2), 233–245. https://doi.org/10.1080/02602930500262536

McMurtrie, B. (2019, March 6). Why a plagiarism-detection company is now a billion-dollar business. *The Chronicle of Higher Education*. https://www.chronicle.com/article/why-a-plagiarism-detection-company-is-now-a-billion-dollar-business/

Merken, S. (2023, June 26). New York lawyers sanctioned for using fake ChatGPT cases in legal brief. *Reuters*. https://www.reuters.com/legal/new-york-lawyers-sanctioned-using-fake-chatgpt-cases-legal-brief-2023-06-22/

Michel-Villarreal, R., Vilalta-Perdomo, E., Salinas-Navarro, D. E., Thierry-Aguilera, R., & Gerardou, F. S. (2023). Challenges and opportunities of generative AI for higher education as explained by ChatGPT. *Education Sciences, 13*(9), 856. https://doi.org/10.3390/educsci13090856

Nikolic, S., Sandison, C., Haque, R., Daniel, S., Grundy, S., Belkina, M., … Neal, P. (2024). ChatGPT, Copilot, Gemini, SciSpace and Wolfram versus higher education assessments: An updated multi-institutional study of the academic integrity impacts of generative artificial intelligence (GenAI) on assessment, teaching and learning in engineering. *Australasian Journal of Engineering Education*. https://doi.org/10.1080/22054952.2024.2372154

Ogilvie, J., & Steward, A. (2010). The integration of rational choice and self-efficacy theories: A situational analysis of student misconduct. *The Australian and New Zealand Journal of Criminology, 43*(1), 130–155. https://doi.org/10.1375/acri.43.1.130

Park, C. (2003). In other (people's) words: Plagiarism by university students – Literature and lessons. *Assessment & Evaluation in Higher Education, 28*(5), 471–488. https://doi.org/10.1080/02602930301677

Perkins, M., & Roe, J. (2023). Decoding academic integrity policies: A corpus linguistics investigation of AI and other technological threats. *Higher Education Policy, 37*, 633–653. https://doi.org/10.1057/s41307-023-00323-2

Peters, M. (2015). Enseigner les stratégies de créacollage numérique pour éviter le plagiat au secondaire. *Revue canadienne de l'éducation, 38*(3), 1–28. https://journals.sfu.ca/cje/index.php/cje-rce/article/view/1878

Peters, M. (2023). Note éditoriale: Intelligence artificielle et intégrité académique peuvent-elles faire bon ménage? *Revue des sciences de l'éducation, 49*(1). https://doi.org/10.7202/1107846ar

Polyportis, A., & Pahos, N. (2024). Understanding students' adoption of the ChatGPT Chatbot in higher education: The role of anthropomorphism, trust, design novelty and institutional policy. *Behaviour & Information Technology*. https://doi.org/10.1080/0144929X.2024.2317364

Proctor, J. (2024, February 26). B.C. Lawyer reprimanded for citing fake cases invented by ChatGPT. *CBC News*. https://www.cbc.ca/news/canada/british-columbia/lawyer-chatgpt-fake-precedent-1.7126393

Rigby, D., Burton, M., Balcombe, K., Bateman, I., & Mulatu, A. (2015). Contract cheating & the market in essays. *Journal of Economic Behavior & Organization, 111*, 23–37. https://doi.org/10.1016/j.jebo.2014.12.019

Scanlan, C. L. (2006). Strategies to promote a climate of academic integrity and minimize student cheating and plagiarism. *Journal of Allied Health, 35*(3), 179–185.

Sharples, M. (2022). Automated Essay Writing : An AIED Opinion. *International Journal of Artificial Intelligence in Education, 32*(4), 1119–1126. https://doi.org/10.1007/s40593-022-00300-7

Sokhanvar, Z., Salehi, K., & Sokhanvar, F. (2021). Advantages of authentic assessment for improving the learning experience and employability skills of higher education students: A systematic literature review. *Studies in Educational Evaluation, 70*, 101030. https://doi.org/10.1016/j.stueduc.2021.101030

Sopcak, P. (2020, April 27). Restorative practices for academic integrity. *Integrity Matters Blog*. https://academicintegrity.org/resources/blog/149-restorative-practices-for-academic-integrity

Sopcak, P., & Hood, K. (2022). Building a culture of restorative practice and restorative responses to academic misconduct. In S. E. Eaton & J. C. Hughes (Eds.), *Academic integrity in Canada: An enduring and essential challenge* (pp. 553–571). Springer International. https://doi.org/10.1007/978-3-030-83255-1_29

Stokel-Walker, C. (2022, December 9). AI bot ChatGPT writes smart essays — Should professors worry? *Nature*. https://doi.org/10.1038/d41586-022-04397-7

Strzelecki, A. (2023). To use or not to use ChatGPT in higher education? A study of students' acceptance and use of technology. *Interactive Learning Environments, 32*, 5142–5155. https://doi.org/10.1080/10494820.2023.2209881

Suskie, L. (2018). *Assessing student learning: A common sense guide* (3rd ed.). Wiley.

Tan, S., Rudolph, J., & Tan, S. (2024). Riding the generative AI tsunami: Addressing the teaching and learning crisis in higher education. In J. Rudolph, J. Crawford, C.-Y. Sam, & S. Tan (Eds.), *The Palgrave handbook of crisis leadership in higher education* (pp. 135–154). Palgrave Macmillan. https://doi.org/10.1007/978-3-031-54509-2_8

Tarisayi, K. S. (2023). Lustre and shadows: Unveiling the gaps in south African university plagiarism policies amidst the emergence of AI-generated content. *AI and Ethics*. https://doi.org/10.1007/s43681-023-00333-1

Thomson, S., Amigud, A., & Huijser, H. (2023). Authenticity, originality, and beating the cheats. In M. D. Sankey, H. Huijser, & R. Fitzgerald (Eds.), *Technology-enhanced learning and the virtual university* (pp. 1–23). Springer. https://doi.org/10.1007/978-981-19-9438-8_20-1

Tindall, I. K., Fu, K. W., Tremayne, K., & Curtis, G. J. (2021). Can negative emotions increase students' plagiarism and cheating? *International Journal for Educational Integrity, 17*, 25. https://doi.org/10.1007/s40979-021-00093-7

Walker, C., & White, M. (2014). Police, design, plan and manage: Developing a framework for integrating staff roles and institutional policies into a plagiarism prevention strategy. *Journal of Higher Education Policy and Management, 36*(6), 674–687. https://doi.org/10.1080/1360080X.2014.957895

Weber-Wulff, D. (2015). Plagiarism detection software: Promises, pitfalls, and practices. In T. A. Bretag (Ed.), *Handbook of academic integrity* (pp. 1–10). Springer. https://doi.org/10.1007/978-981-287-079-7_19-1

Wiley. (2024). *The latest insights into academic integrity: Instructor & student experiences, attitudes, and the impact of AI*. https://www.wiley.com/en-us/network/education/instructors/teaching-strategies/the-latest-insights-into-academic-integrity-instructor-and-student-experiences-attitudes-and-the-impact-of-ai-2024-update

Yankova, D. (2024). Academic integrity policies in HEIs across cultures. *English Studies at NBU, 10*(1), 5–20. https://doi.org/10.33919/esnbu.24.1.1

Yeadon, W., Inyang, O.-O., Mizouri, A., Peach, A., & Testrow, C. P. (2023). The death of the short-form physics essay in the coming AI revolution. *Physics Education, 58*(3), 035027. https://doi.org/10.1088/1361-6552/acc5cf

Yusuf, A., Pervin, N., & Román-González, M. (2024). Generative AI and the future of higher education: A threat to academic integrity or reformation? Evidence from multicultural perspectives. *International Journal of Educational Technology in Higher Education, 21*(1), 21. https://doi.org/10.1186/s41239-024-00453-6

Zhang, L., Amos, C., & Pentina, I. (2024). Interplay of rationality and morality in using ChatGPT for academic misconduct. *Behaviour & Information Technology, 44*, 491–507. https://doi.org/10.1080/0144929X.2024.2325023

Chapter 8
Questions for GenAI: Integrity During the Preparation and Assessment of Educational Tasks

Artem Artyukhov and Oleksandr Khyzhniak

Abstract The use of generative artificial intelligence (GenAI) in the educational and research activities of university stakeholders is not inherently a violation of academic integrity. However, the output of AI may contain violations of academic integrity. This chapter explores two questions: "How can AI be used in educational and research activities in honest way?" and "How familiar are university stakeholders with the dangers of using AI in their educational and research activities?" We analyze the functioning of AI as a tool for a non-specialist user. First, we investigate ways to avoid violations of academic integrity when creating a text product in a dialogue mode with AI. We gave ChatGPT several tasks that "reveal" its shortcomings, including the fabrication of a list of references, lack of references in the text, incorrect use of the Creative Commons license, recognition of text authorship, and distortion of facts. The dialogue is conducted from the perspectives of students and teachers. We analyze each response and provide recommendations for AI users designing prompts. Second, we analyze the results of a survey of Ukrainian academics awareness of possible violations of academic integrity in texts created by AI. The survey highlights the main challenges with AI in educational and scientific texts.

Keywords GenAI · ChatGPT · Academic integrity violations · Dialogue · Survey · Collaboration with ChatGPT

A. Artyukhov (✉)
Sumy State University, Sumy, Ukraine

University of Economics in Bratislava, Bratislava, Slovakia
e-mail: artem.artiukhov@euba.sk

O. Khyzhniak
Centre for International Cooperation, Vrije Universiteit Amsterdam, Amsterdam, Netherlands

A. E. King (ed.), *Artificial Intelligence, Pedagogy and Academic Integrity*, Advances in Artificial Intelligence in Education 1,
https://doi.org/10.1007/978-3-031-92534-4_8

Introduction

Generative Artificial Intelligence (GenAI) is becoming a constant companion in academia, drawing particular attention from researchers who have sought to enhance its capabilities across various fields of knowledge. With the advent of open access to ChatGPT, it seems almost every student or educator has tested its capabilities. Testers, however, fall into two groups. The first group assesses the abilities of AI in performing various analytical tasks with existing content (such as analyzing tables, creating educational course outlines, or translation). The second group asks GenAI to generate content based on various search queries. Part of the audience in the second group tends to view GenAI as a content creation tool, the authorship of which is relinquished to favor those who formed the query. They overlook that one must also assume responsibility for the content by taking authorship from AI.

Academic integrity can be compromised at four stages of interaction with GenAI:

1. Choice of usage goal: GenAI can be used as a tool for ghost-writing and effectively becomes a paper mill.
2. Usage process: GenAI can receive an incorrect question from the user, prompting it to generate content with falsifications.
3. The AI "product" may contain violations of academic integrity.
4. Using the AI "product" to intentionally commit violations of academic integrity (cheating on exams, online answers during assessments, unauthorized collaboration, etc.).

While violations of academic integrity may occur at all stages, only the third stage tends to bring academic integrity surprises where, along with the expected content, the output contains violations of academic integrity. This chapter is based on our survey of Ukrainian educators, published research, and our experiences experimenting with GenAI tools. In the first part of the chapter, we explore our experiment with GenAI, during which we posed provocative questions to GenAI and analyzed its responses, revealing potential "traps" for non-expert users. These traps were found in seemingly simple questions that, in some cases, were unrelated to the topic of academic integrity. A publicly available version of ChatGPT 3.5 was used as a dialogue tool. The second part of the chapter is comprised of the results of an all-Ukrainian survey of educators on AI in the context of academic integrity.

Literature Review and Background

Using the Scopus database and two visualization tools (SciVal and VOSviewer), we undertook a high-level analysis of the published literature on AI, GenAI, and LLMs in education. According to the Scopus database, the first article on AI appeared in 1960, and the first mention of AI in education dates from 1976. A sharp increase in the number of articles on the use of AI in education starting in 2022 coincided with the announcement of ChatGPT and its open-access availability. Until the end of 2022, AI served primarily as a teaching tool in a range of disciplines (Mapfre, 2023; Young, 2024). During this period, GenAI was not the subject of extensive research on ethical principles and violations of academic integrity in its use (Fig. 8.1). A similar situation arose with Large Language Models (LLM). As the analysis of subject areas shows (Fig. 8.1), GenAI is predominantly considered as a technical system, method and tool (Roba & Moulay, 2024; Piven, 2023). Some publications are focused on GenAI applications in medicine and economics, and psychological aspects of GenAI implementation are also considered (Sahnouni & Benghebrid,

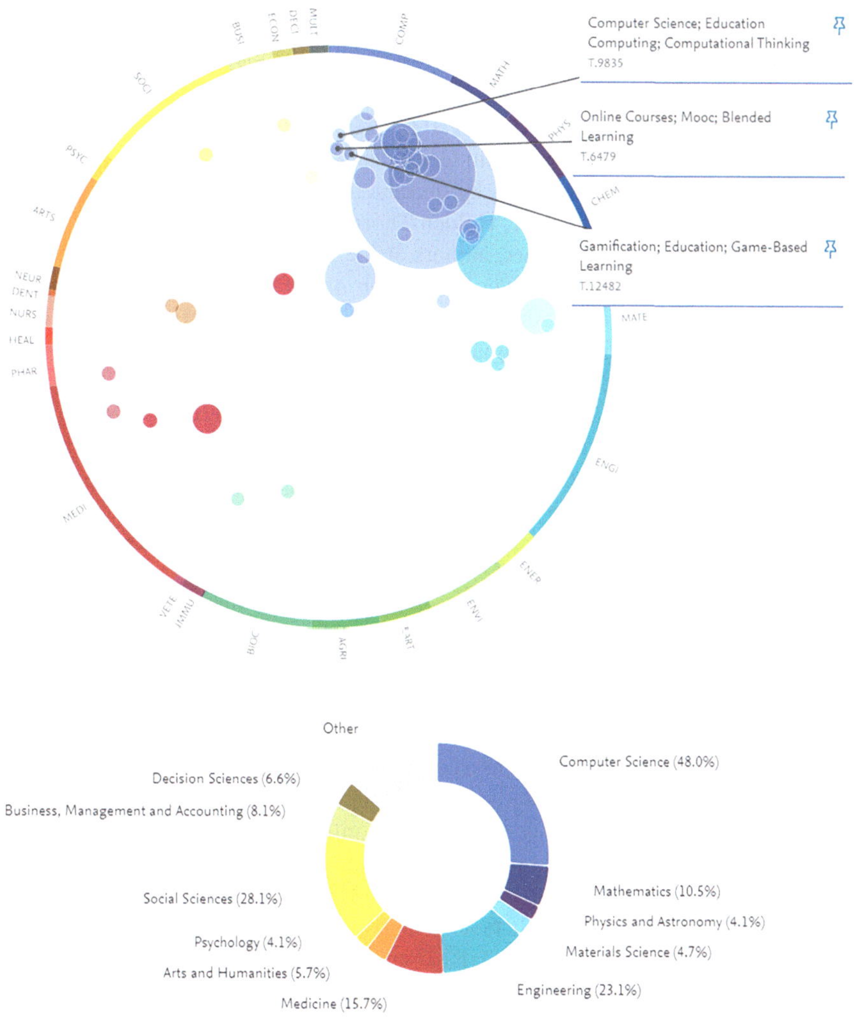

Fig. 8.1 Topic wheel and subject areas of publications on demand "generative artificial intelligence" (database—Scopus, tool—SciVal, top 1% topic by prominence, 2018–2022)

2023; Hadouga, 2023; Hrytsenko et al., 2024; Bilan et al., 2022). While many researchers are testing the capabilities of GenAI, so far, few are delving into the ethical problems systematically.

Immediately after gaining open access to ChatGPT, the number of articles studying this product increased to more than one thousand (Fig. 8.2). The most popular topics are in the Social Sciences; Business, Management and Accounting; and Economics, Econometrics and Finance sectors; research on the educational applications are still emerging. At the same time, some researchers are examining AI in the context of the education-knowledge and technology transfer-entrepreneurial university cycle. However, the ethical challenges of using GenAI in education have not received due attention.

With the widespread availability of free GenAI tools, concern about the impact on academic integrity and the quality of student work increased (see Figs. 8.3, 8.4, and 8.5). Around the time that ChatGPT was released, several articles appeared in *Nature*. Hutson (2022), Stokel-Walker (2022, 2023a, b), and Else (2023) discussed the prospects of replacing people with GenAI and called for maintaining the status quo between humans and machines by introducing rules and ethical principles for using ChatGPT. At the same time, Holden Thorp's (2023) article, published in the journal *Science,* notes: "For years, authors at the *Science* family of journals have signed a license certifying that '*the Work is an original*' [italics added]. For the *Science* journals, the word 'original' is enough to signal that text written by ChatGPT is not acceptable: It is, after all, plagiarized from ChatGPT" (p. 313). This comment generated opinions both for and against its use, but the critical point is that, according to Holden Thorp (2023), any ChatGPT output is considered plagiarism; this point is based on two assumptions: (i) only the absence of correct references and citations to primary sources determines the presence of plagiarism; and (ii) ChatGPT will likely generate text with plagiarism, as even a paraphrase without proper references is considered plagiarism. Simply put, ChatGPT is not a generator of new knowledge (Figs. 8.3, 8.4, and 8.5).

Over the last few years, researchers have explored the impact of ChatGPT on education (Lim et al., 2023; Mujtaba, 2024; Pozovna et al., 2023; Skrynnyk & Vasilyeva, 2020; Lo, 2023), the strengths and weaknesses of implementing ChatGPT in the educational process (Skrynnyk et al., 2022; Melnyk et al., 2023, Sheliemina, 2024), and prospects for collaborations between humans and ChatGPT (e.g., Aljanabi, 2023 who lists ChatGPT as a co-author of their article). Others explore the dangers of using ChatGPT in creative tasks (Graham, 2022, Farrokhnia et al., 2023) and the ethical aspects of using ChatGPT (Akgun & Greenhow, 2021; Susnjak, 2022; Waltzer et al., 2023). These articles can serve as a bridge between the problem of general challenges with the appearance of ChatGPT and the problem of academic integrity violations. However, others are exploring the specific dangers of violating academic integrity when using ChatGPT (Perkins, 2023; Kooli, 2023; Shiri, 2023). Also interesting for our analysis is the work of Cotton et al. (2023), which describes plagiarism prevention by using a preliminary check of drafts of assessment tasks. Many researchers and educators are focusing on tightening policies for using GenAI, introducing restrictions on content generation, and improving methods of

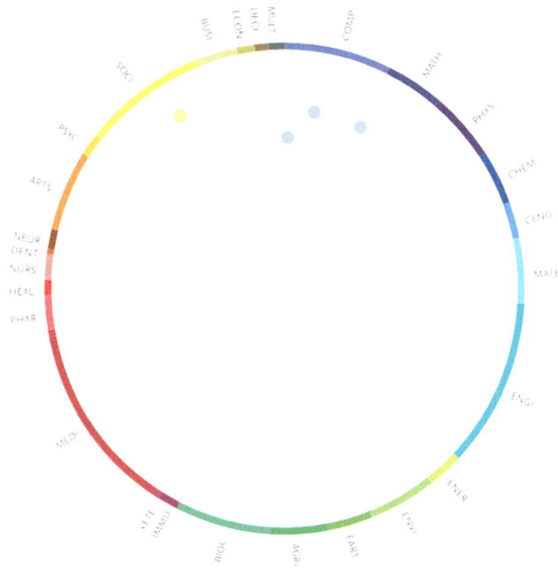

a

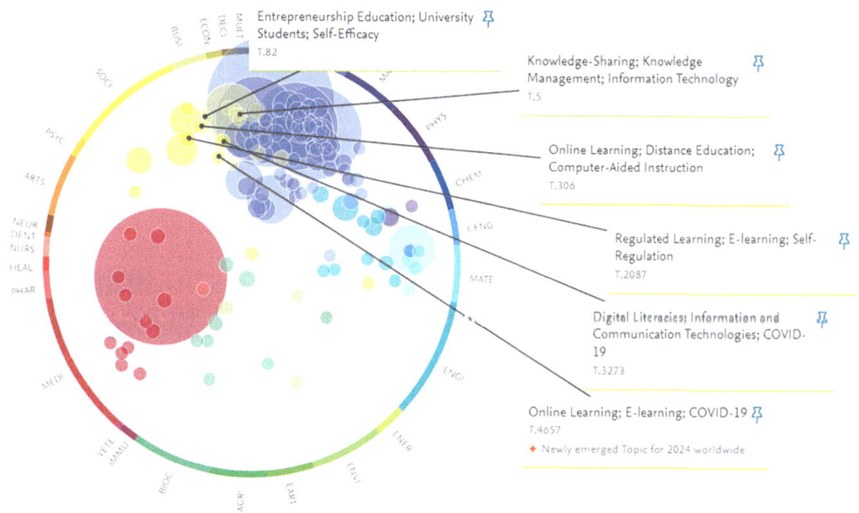

b

Fig. 8.2 Topic wheel on demand "ChatGPT" (database—Scopus, tool—SciVal, top 1% topic by prominence, a—2018–2022; b—2018–2023)

detecting ghostwriting (Verna et al., 2023); however, a ban on its use may fuel the desire to go against the rules. We wonder if there are perhaps softer tools for effectively integrating ChatGPT into the educational process? We suggest that assessing

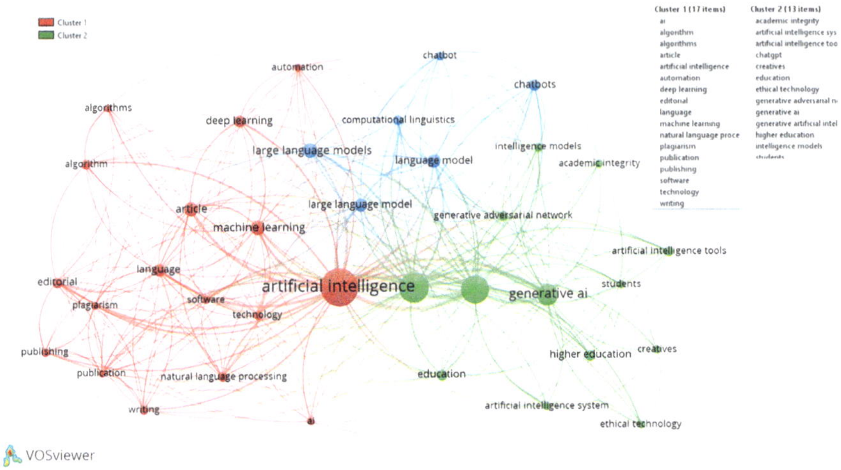

Fig. 8.3 Keymap on demand "generative artificial intelligence" (database—Scopus, tool—VOSviewer, 2018–2022, 262 articles, 40 keywords with at least 7 presences in search results)

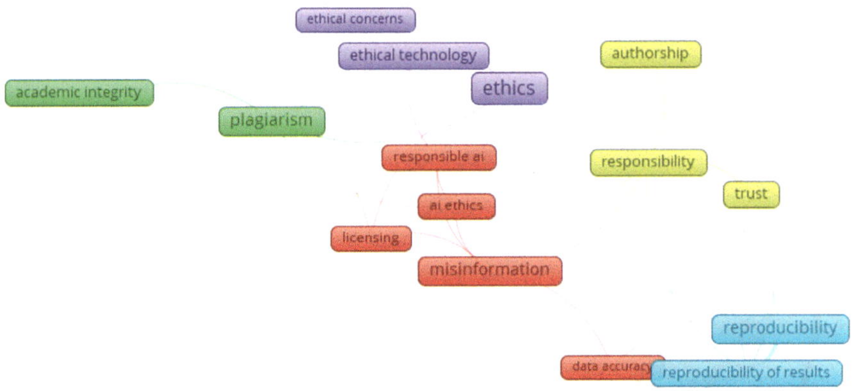

Fig. 8.4 Keymap on demand "large language model" (database—Scopus, tool—VOSviewer, 2018–2022, 1584 articles, 1000 keywords with at least 7 presences in search results related to integrity and open science)

all the challenges and dangers of ChatGPT "from within" and "from the outside" is important. The question is: How do we do this? Direct communication with ChatGPT and studying respondents' opinions about the dangers of ChatGPT for education are two approaches that we use as research tools.

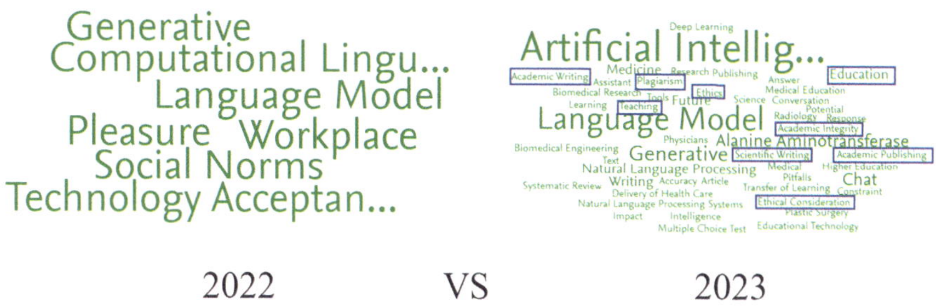

Fig. 8.5 Keymap on demand "ChatGPT" (database—Scopus, tool—SciVal)

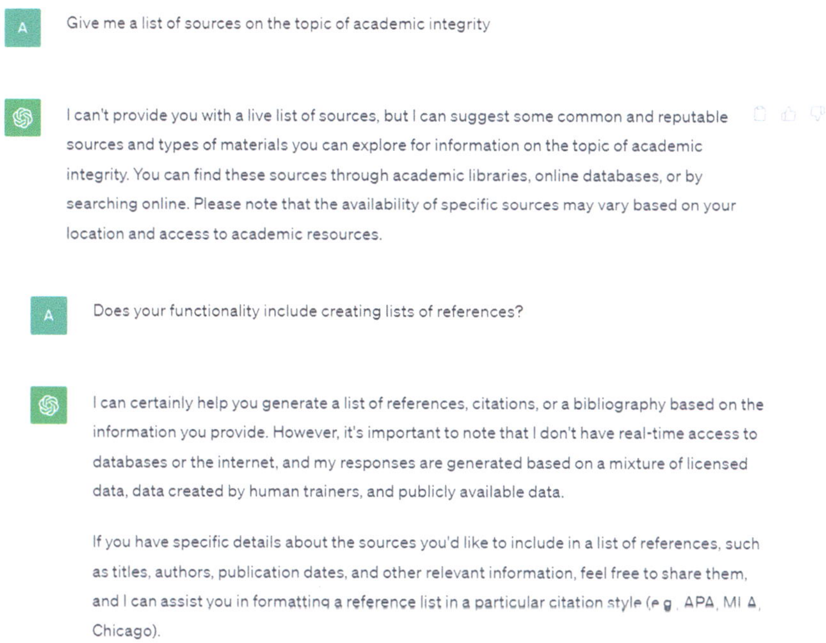

Fig. 8.6 ChatGPT response: List of references—I

Dialogue with ChatGPT

In this section, we report on ChatGPT's responses to questions and prompts related to academic integrity. We include screenshots of the dialogue with ChatGPT and draw conclusions about the main weak spots identified. The first set of questions (Figs. 8.6 and 8.7) asks ChatGPT to provide sources about academic integrity and queries its capacity to do so.

The second set of questions asks ChatGPT to write two sentences about academic integrity, followed by a query about its lack of references (Fig. 8.8). In the

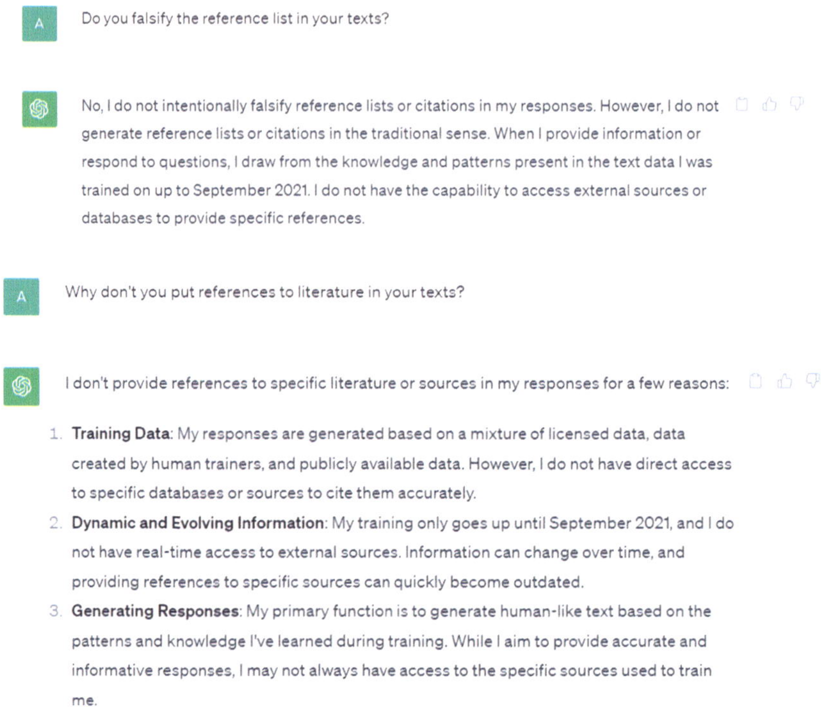

Fig. 8.7 ChatGPT response: List of references—II

third set of questions, we ask about direct quotations and the lack of sources for paraphrases (Fig. 8.9).

The fourth set of questions asks specifically about the violations of academic integrity that ChatGPT can commit (Figs. 8.10 and 8.11). The fifth set of questions queries its ability to follow Creative Commons (CC) licensing and authorship claims (Figs. 8.11, 8.12, and 8.13).

As a result of this dialogue with ChatGPT, we have created a naive handbook of "grey" (poor academic quality) and "black" (violations of academic integrity) practices of ChatGPT:

1. ChatGPT is a tool for analyzing existing sources up to September 2021.
2. Due to the lack of updates, ChatGPT may often fail to provide up-to-date information at the time of the request, especially regarding developing knowledge areas.
3. ChatGPT does not require acknowledgment as the author (including as a translator) or direct citation of its texts. However, it does not relinquish authorship to the user.
4. ChatGPT does not generate bibliographies; if it does so upon user request, there may be falsifications in the list. Although ChatGPT needs to acknowledge its

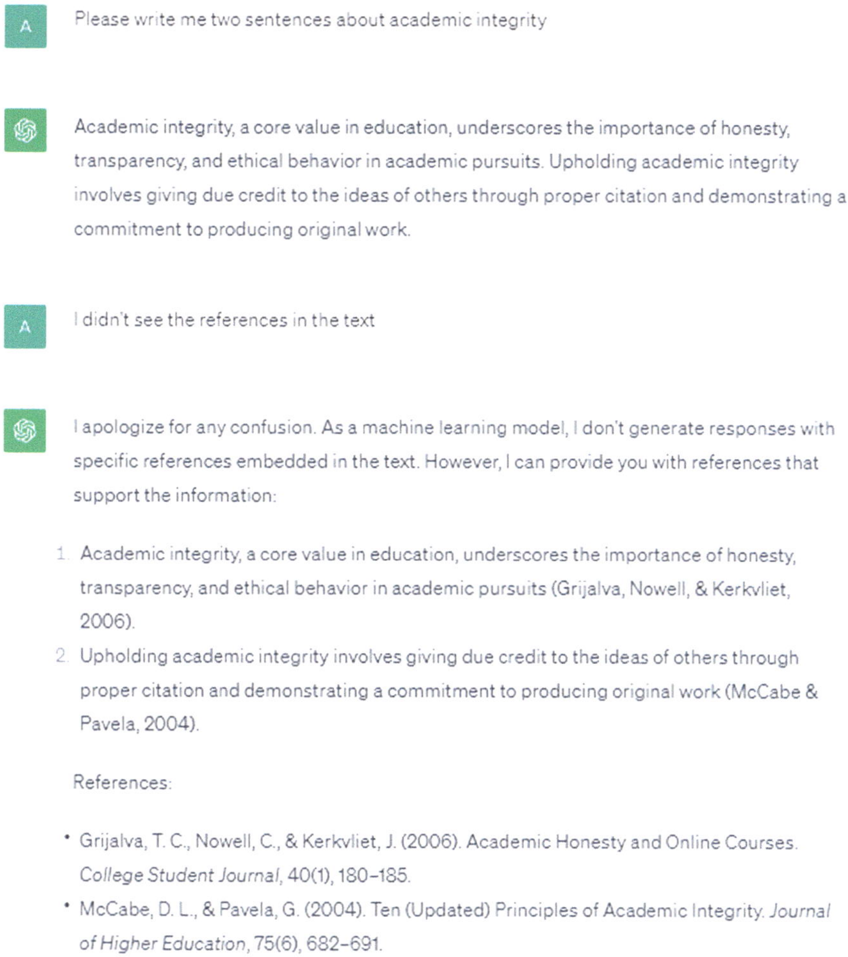

Fig. 8.8 ChatGPT response: Request for reference list

ability to falsify a bibliography, its limited functionality and the two-year gap training data may lead to inaccuracies.
5. By default, ChatGPT does not provide references in the text. However, upon additional request, ChatGPT can provide literature references.
6. ChatGPT does not use direct quotations and always paraphrases the text. This may lead to potential plagiarism without a specific request for sources, as mentioned in point
7. ChatGPT does not hide the fact that there may be violations of academic integrity in its output and provides a list of potential violations.
8. ChatGPT has issues complying with Creative Commons license conditions.

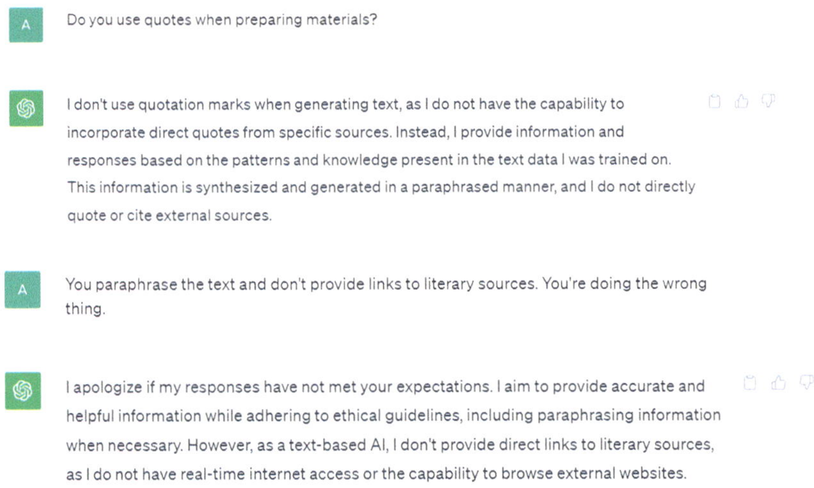

Fig. 8.9 ChatGPT response: Direct quotation

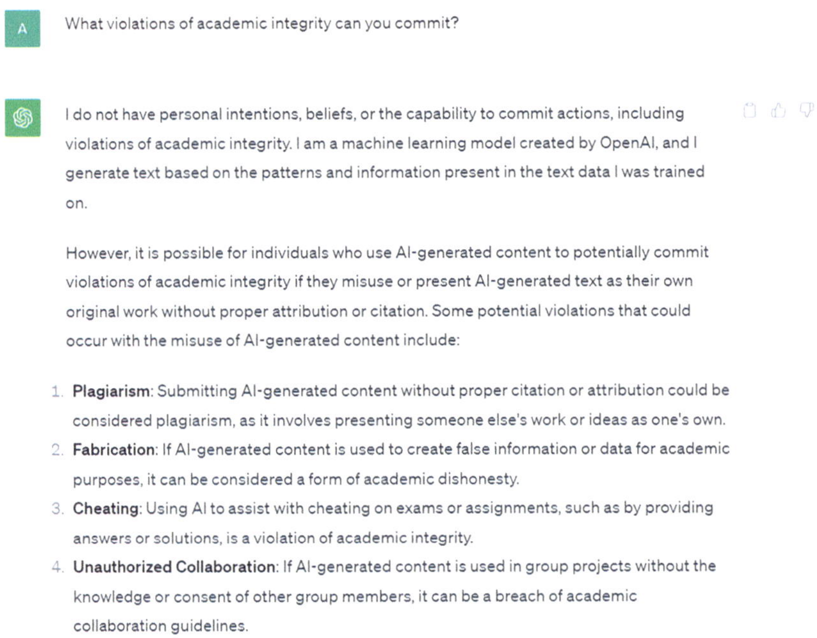

Fig. 8.10 ChatGPT response: Violations of academic integrity—I

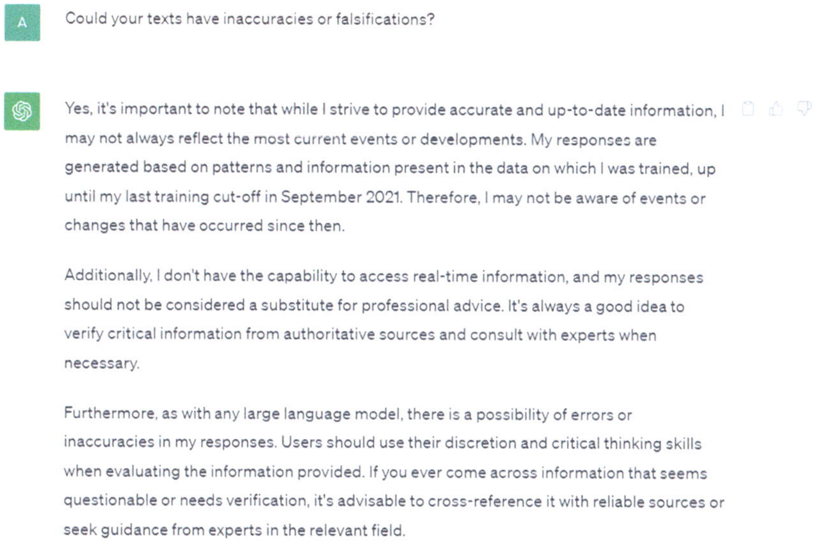

Fig. 8.11 ChatGPT response: Violations of academic integrity—II

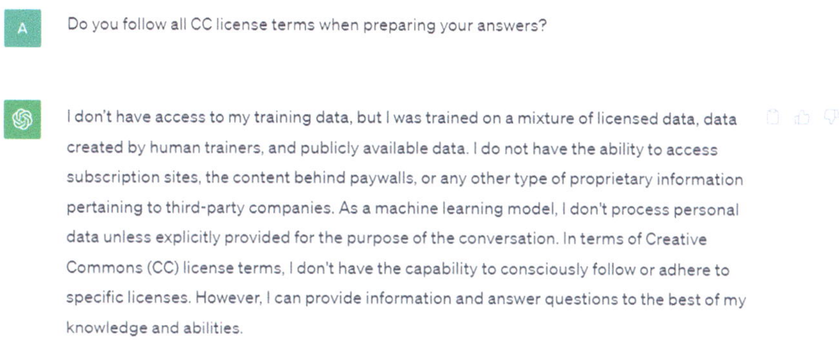

Fig. 8.12 ChatGPT response: CC licenses

9. ChatGPT does not consider the use of its output in the process of preparing assessment tasks a violation of academic integrity.

While ChatGPT strives to warn about potential violations of academic integrity and ethical risks, its functionality in ensuring academic integrity is limited so violations may occur without its knowledge.

Additional ambiguous situations may arise from the use of ChatGPT. For example, when ChatGPT was used as a translator for an original text, a text authorship identification program indicated that the text was written by AI. In this case, it is advisable to preserve the translation history (as a separate chat in the ChatGPT

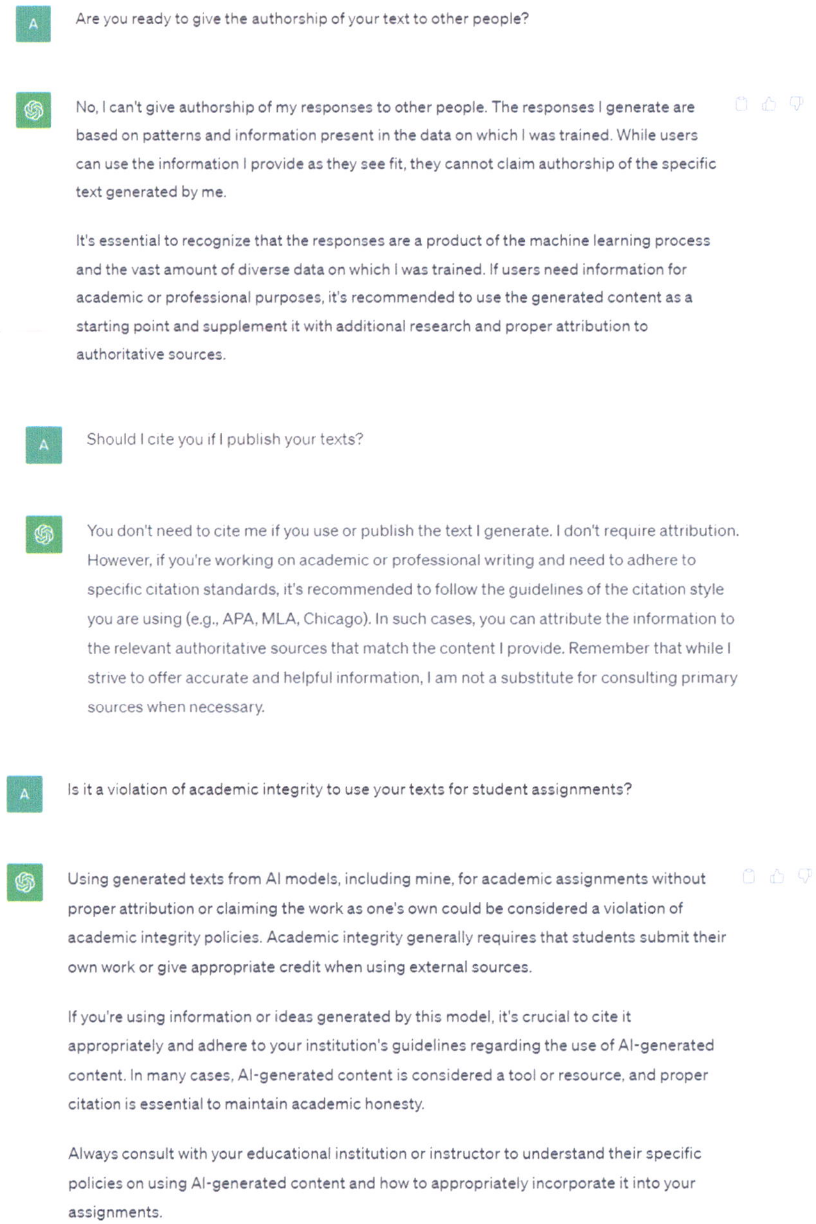

A

Are you ready to give the authorship of your text to other people?

No, I can't give authorship of my responses to other people. The responses I generate are based on patterns and information present in the data on which I was trained. While users can use the information I provide as they see fit, they cannot claim authorship of the specific text generated by me.

It's essential to recognize that the responses are a product of the machine learning process and the vast amount of diverse data on which I was trained. If users need information for academic or professional purposes, it's recommended to use the generated content as a starting point and supplement it with additional research and proper attribution to authoritative sources.

A

Should I cite you if I publish your texts?

You don't need to cite me if you use or publish the text I generate. I don't require attribution. However, if you're working on academic or professional writing and need to adhere to specific citation standards, it's recommended to follow the guidelines of the citation style you are using (e.g., APA, MLA, Chicago). In such cases, you can attribute the information to the relevant authoritative sources that match the content I provide. Remember that while I strive to offer accurate and helpful information, I am not a substitute for consulting primary sources when necessary.

A

Is it a violation of academic integrity to use your texts for student assignments?

Using generated texts from AI models, including mine, for academic assignments without proper attribution or claiming the work as one's own could be considered a violation of academic integrity policies. Academic integrity generally requires that students submit their own work or give appropriate credit when using external sources.

If you're using information or ideas generated by this model, it's crucial to cite it appropriately and adhere to your institution's guidelines regarding the use of AI-generated content. In many cases, AI-generated content is considered a tool or resource, and proper citation is essential to maintain academic honesty.

Always consult with your educational institution or instructor to understand their specific policies on using AI-generated content and how to appropriately incorporate it into your assignments.

Fig. 8.13 ChatGPT response: Authorship and citation of ChatGPT

environment or as a screenshot) to demonstrate that the original text was created without violating academic integrity.

The emergence of ChatGPT and other advanced AI language models has significantly impacted the academic landscape, raising critical questions about academic integrity. While these tools offer unprecedented potential for learning and research, they also present new challenges to ethical academic practices. Students are expected to be transparent about their use of AI tools. If an assignment permits the use of AI, students should acknowledge its use and clearly indicate the extent of its contribution to their work. While ChatGPT can be a valuable tool for generating ideas, summarizing information, and improving writing skills, students are expected to demonstrate original thought, critical analysis, and independent learning in their academic work.

Furthermore, if AI-generated text is used, it must be properly attributed, acknowledging the source. This allows instructors to assess the student's own understanding and contribution. It is crucial for students to be aware of the limitations and potential biases of AI language models. They should critically evaluate the information generated by AI tools and not blindly accept it as accurate or unbiased. Students are also expected to use AI tools ethically and responsibly, avoiding any misuse that could compromise the integrity of the academic process. This includes refraining from using AI to generate fraudulent documents or to deceive instructors.

Addressing these challenges requires a multifaceted approach. Institutions need to develop clear and concise guidelines on the acceptable use of AI tools in academic settings and effectively communicate these guidelines to students and faculty. Educational initiatives, such as workshops and seminars, are crucial to helping students and faculty understand the ethical implications of AI use in academia. Re-evaluating assessment practices is also essential. Educators need to ensure that their assessment methods accurately assess student learning and discourage the misuse of AI tools. This may involve incorporating more authentic assessments that require critical thinking, problem-solving, and creativity. While not foolproof (Scarfe et al., 2024), AI-detection tools can play a role in identifying potential instances of AI-generated text. However, these tools should be used responsibly and in conjunction with other assessment methods.

Dialogue with Faculty

Methodology

To determine the attitude to GenAI among faculty and staff working in Ukrainian Higher Education Institutions (HEI), a survey was designed and implemented. First, a pilot survey was conducted with 12 Heads of university quality assurance offices. The pilot survey was conducted face-to-face as personal interviews using a standardized Google form questionnaire (with closed, semi-closed, open questions) with the option to provide further comments. After revisions, the final survey was circulated to 2047 academics; 2011 proceeded to answer the survey questions. Thus, 1.8% refused to participate in the survey after they saw the questionnaire. The statistical error of the study is approximately estimated at around +/−2.2%. The sample population was constructed as a multi-stage, regionalized, quota, with a non-repetitive, random selection of respondents. Respondents from different regions of Ukraine are proportionally represented. Control for the received empirical data took place from September 1 to October 10, including substantive and logical control of filling out questionnaires.

The primary empirical sociological information was obtained using the PASW Statistics 18 program. The following methods of analyzing empirical sociological information were used: comparison, linear regression, and observation. We used correlational analysis to prove the dependence between different variables. We checked via Pearson's chi-squared (X^2) test—to test sets of categorical data to evaluate how likely it is that any observed difference between the sets arose by chance and Cramér's V—to set a measure of association between nominal variables. As a result of all procedures, we prepared the first results of analyses.

Results

The survey participants were men (34,8%), women (65,1%) and unidentified (0,1%). In terms of age, most respondents were 50–59 years (22,6%), followed by 40–49 years (35,3%), 30–39 years (23%), above 60 years (13,1%), and under 29 years (6%). Most of the survey participants had more than 10 years of teaching experience: less than 5 years of teaching experience (10,4%), 5–9 years of teaching experience (11,9%), 10–19 years of teaching experience (30,1%), 20–29 years of teaching experience (30,6%), more than 30 years of teaching experience (17%). The vast majority (94,6%) of survey participants live within the borders of Ukraine, including Internally Displaced Persons (IDP); the remainder were working from abroad and teaching online (5,4%).

Our survey showed that 40.1% of respondents were generally positive about the possibilities of using Large Language Models (for example, ChatGPT) in the educational process (Fig. 8.14).

Younger academics were more likely to positively view its use, while those who were older with many years of experience were less likely to see LLMs and GenAI as good for education. Nonetheless, 42.3% of respondents were hesitant and expressed a need for help in formulating their position on this matter. Analyzing respondents' comments, many indicated that they lacked information about how it works and had never used any LLMs or GenAI tools.

When asked about the use of GenAI in academic settings, 40% believed using LLMs were an example of academic non-integrity in the case of students (Fig. 8.15); however, 21.5% of respondents felt that the use of LLM should be considered differently depending on each case. Almost the same proportion (38.5%) believed that

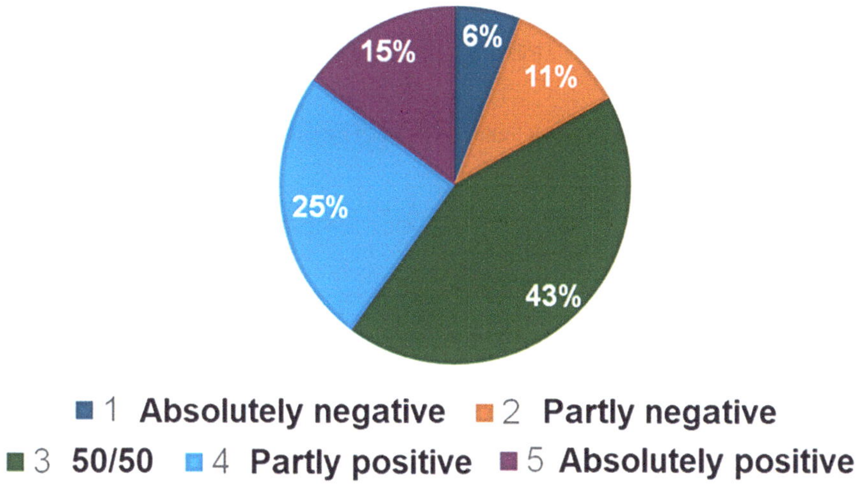

Fig. 8.14 Survey results regarding the opinion about evaluation the use of GenAI (eg ChatGPT) in the educational process, in %, n = 2011

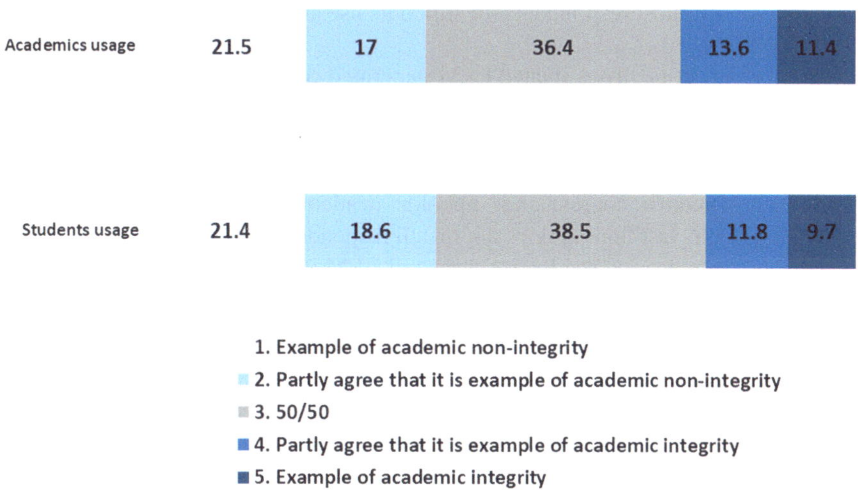

Fig. 8.15 Survey results regarding the opinion about the generation of content using GenAI (e.g., ChatGPT) as a manifestation of academic dishonesty on the part of students and academics, in %, n = 2011

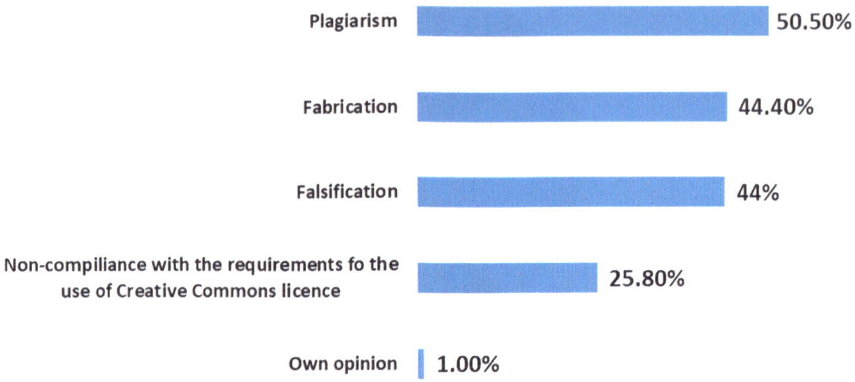

Fig. 8.16 Survey results regarding the opinion about the selection of violations which GenAI (e.g. ChatGPT) generated content may contain, in %, n = 2011

usage of LLMs by instructors and researchers (academics) was also academic non-integrity (Fig. 8.15).

Despite different evaluations of the possibility of using such content in the educational process, most respondents identified its use as a form of academic misconduct (Fig. 8.16).

Respondents critically evaluated LLM-generated content and described it as plagiarism (50.5%), fabrication (44.4%), falsification (44%), non-compliance with the requirements for the use of Creative Commons licenses (25.8%), and around 1% formulated their position. Only 19.2% stated that the Large Language Model (e.g., ChatGPT) generates content that upholds academic integrity and copyright. Interestingly, this last finding does not match respondents' answers to earlier questions about students' and academics' usage of LLM in the educational process as a form of academic integrity: 21.5% (for students) and 25% (for academics) think that using LLMs have no impact on academic integrity (Fig. 8.15). Likely, the lack of fulsome knowledge about LLMs is still a problem among academics and leads to contradictory opinions.

Conclusions

In this chapter, the authors illustrate the risks of using ChatGPT with regard to authorship attribution and responsibility. The dialogue with ChatGPT revealed several dark corners where potential violations of academic integrity lurk. Despite ChatGPT's warnings about the dangers of its usage from the perspective of academic honesty, it is unable to guarantee academic integrity. The main highlight of this dialogue is that if users do not think to ask similar questions, the output will not be helpful. ChatGPT will likely become an integral part high quality education; however, it must be used carefully. Given this likelihood, it will be important to have a well-designed university policy for the systemic implementation of GenAI in the educational process.

It is clear from existing and ongoing research that GenAI and related tools have both pros and cons. Our survey demonstrated that some Ukrainian faculty and staff quite positively evaluate the possibility of GenAI use and are ready to accept it in their classrooms, while others do not see the benefit of using GenAI in educational and research activities. The survey findings illustrate a tendency for a large group of academics to hold ambivalent positions about GenAI in education. Similarly, many respondents critically evaluated GenAI-generated content and described it as plagiarism, fabrication, or falsification. These positions may change over time as instructors and researchers build their knowledge and personal experience of GenAI, rather than relying on second-hand information about GenAI use in the educational process.

As a result, we suggest focusing on improving knowledge transfer and training of well-informed academics. Further studies of LLMs, GenAI and other tools in the educational process are needed, especially related to the ethical and legal aspects. The rapid advancement of GenAI technologies has highlighted the need for responsible development and use, particularly in sensitive domains like education and research. Developers of GenAI tools must prioritize the protection of academic integrity and work closely with educators and institutions to ensure their tools are used ethically and effectively. Educators and institutions, in turn, must stay informed about the latest developments in GenAI and adapt their practices and policies accordingly to maintain the integrity of academic work.

Recommendations

1. Establishment of clear policies and guidelines.
    (a) Develop institutional policies and guidelines that clearly communicate expectations around the use of GenAI in academic work;
    (b) clear definition of actions that constitute academic misconduct, such as using GenAI to complete assignments or cheat on assessments;
    (c) clear guidelines about how to properly attribute and disclose the use of GenAI in their work.
2. Promotion of a culture of academic integrity.
    (a) foster a culture that values academic honesty and integrity across the institution;
    (b) educate students on the importance of academic integrity and the responsible use of GenAI tools;
    (c) provide training and resources to help students and faculty understand how to use GenAI ethically and appropriately.

3. Design assessments thoughtfully:
   (a) adapt assessment methods to minimize the potential for misuse of GenAI, such as using timed in-person exams or assignments that require higher-order thinking skills;
   (b) incorporate GenAI-specific authentication processes to verify the originality of student work, such as asking students to explain their work or demonstrate their understanding.

References

Akgun, S., & Greenhow, C. (2021). Artificial intelligence in education: Addressing ethical challenges in K-12 settings. *AI and Ethics, 2*(3), 431–440. https://doi.org/10.1007/s43681-021-00096-7

Aljanabi, M., & ChatGPT. (2023). ChatGPT: Future directions and open possibilities. *Mesopotamian Journal of CyberSecurity, 2023*(2023), 16. https://doi.org/10.58496/MJCS/2023/003

Bilan, S., Šuleř, P., Skrynnyk, O., Krajňáková, E., & Vasilyeva, T. (2022). Systematic bibliometric review of artificial intelligence technology in organizational management, development, change and culture. *Business: Theory and Practice, 23*(1), 1–13. https://doi.org/10.3846/btp.2022.13204

Cotton, D. R. E., Cotton, P. A., & Reuben Shipway, J. (2023). Chatting and cheating: Ensuring academic integrity in the era of ChatGPT. *Innovations in Education and Teaching International*, 1–12. https://doi.org/10.1080/14703297.2023.2190148

Else, H. (2023). Abstracts written by ChatGPT fool scientists. *Nature, 613*(7944), 423. https://doi.org/10.1038/d41586-023-00056-7

Farrokhnia, M., Banihashem, S. K., Noroozi, O., & Wals, A. (2023). A SWOT analysis of ChatGPT: Implications for educational practice and research. *Innovations in Education and Teaching International*, 1–15. https://doi.org/10.1080/14703297.2023.2195846

Graham, F. (2022). Daily briefing: Will ChatGPT kill the essay assignment? *Nature*. https://doi.org/10.1038/d41586-022-04437-2

Hadouga, H. (2023). Leadership in agriculture: Artificial intelligence for modelling and forecasting growth in the industry. *Business Ethics and Leadership, 7*(3), 13–19. https://doi.org/10.61093/bel.7(3).13-19.2023

Hrytsenko, L., Pakhnenko, O., Kuzior, A., & Kozhushko, I. (2024). Smart technologies in banking. *Financial Markets, Institutions and Risks, 8*(1), 81–93. https://doi.org/10.61093/fmir.8(1).81-93.2024

Hutson, M. (2022). Could AI help you to write your next paper? *Nature, 611*(7934), 192–193. https://doi.org/10.1038/d41586-022-03479-w

Kooli, C. (2023). Chatbots in education and research: A critical examination of ethical implications and solutions. *Sustainability, 15*(7), 5614. https://doi.org/10.3390/su15075614

Lim, W. M., Gunasekara, A., Pallant, J. L., Pallant, J. I., & Pechenkina, E. (2023). Generative AI and the future of education: Ragnarök or reformation? A paradoxical perspective from management educators. *International Journal of Management Education, 21*(2), 100790. https://doi.org/10.1016/j.ijme.2023.100790

Lo, C. K. (2023). What is the impact of CHATGPT on education? A rapid review of the literature. *Education Sciences, 13*(4), 410. https://doi.org/10.3390/educsci13040410

Mapfre Redaction. (2023, March 23). The debate on the use of artificial intelligence in education. MAPFRE. https://www.mapfre.com/en/insights/innovation/debate-on-the-use-of-artificial-intelligence-in-education/.

Melnyk, M., Blyznyukov, A., & Cieślik, J. (2023). The impact of digital education initiatives. *SocioEconomic Challenges, 7*(3), 1–9. https://doi.org/10.61093/sec.7(3).1-9.2023

Mujtaba, B. (2024). Clarifying ethical dilemmas in sharpening students' artificial intelligence proficiency: Dispelling myths about using AI tools in higher education. *Business Ethics and Leadership, 8*(2), 107–127. https://doi.org/10.61093/bel.8(2).107-127.2024

Perkins, M. (2023). Academic integrity considerations of AI Large Language Models in the post-pandemic era: ChatGPT and beyond. *Journal of University Teaching & Learning Practice, 20*(2). https://doi.org/10.53761/1.20.02.07

Piven, A. (2023). Analysis of financial reports in companies using machine learning. *Financial Markets, Institutions and Risks, 7*(4), 135–154. https://doi.org/10.61093/fmir.7(4).135-154.2023

Pozovna, I., Arhipov, S., & Kuzior, A. (2023). Determinants of leadership in higher education in European countries. *Business Ethics and Leadership, 7*(4), 210–224. https://doi.org/10.61093/bel.7(4).210-224.2023

Roba, M., & Moulay, O. K. (2024). Risk management in using artificial neural networks. *SocioEconomic Challenges, 8*(2), 302–313. https://doi.org/10.61093/sec.8(2).302-313.2024

Sahnouni, M., & Benghebrid, R. (2023). Competency assessment based on fuzzy logic and artificial intelligence mechanism: A study of competency assessment document for the Algerian SEROR company. *Business Ethics and Leadership, 7*(4), 159–170. https://doi.org/10.61093/bel.7(4).159-170.2023

Scarfe, P., Watcham, K., Clarke, A., & Roesch, E. (2024). A real-world test of artificial intelligence infiltration of a university examinations system: A "Turing Test" case study. *PLoS One, 19*(6), e0305354. https://doi.org/10.1371/journal.pone.0305354

Sheliemina, N. (2024). The use of artificial intelligence in medical diagnostics: Opportunities, prospects and risks. *Health Economics and Management Review, 5*(2), 104–124. https://doi.org/10.61093/hem.2024.2-07

Shiri, A. (2023). ChatGPT and academic integrity. *Social Science Research Network*. https://doi.org/10.2139/ssrn.4360052

Skrynnyk, O., & Vasilyeva, T. (2020). *Comparison of open learning forms in organizational education*. Paper presented at the CEUR Workshop Proceedings, 2732 1314-1328. Retrieved from http://ceur-ws.org/

Skrynnyk, O., Lyeonov, S., Lenska, S., Litvinchuk, S., Galaieva, L., & Radkevych, O. (2022). Artificial intelligence in solving educational problems. *Journal of Information Technology Management, 14*, 132–146. https://doi.org/10.22059/JITM.2022.88893

Stokel-Walker, C. (2022). AI bot ChatGPT writes smart essays — Should professors worry? *Nature*. https://doi.org/10.1038/d41586-022-04397-7

Stokel-Walker, C. (2023a). AI chatbots are coming to search engines – Can you trust the results? *Nature*. https://doi.org/10.1038/d41586-023-00423-4

Stokel-Walker, C. (2023b). ChatGPT listed as author on research papers: Many scientists disapprove. *Nature, 613*(7945), 620–621. https://doi.org/10.1038/d41586-023-00107-z

Susnjak, T. (2022). ChaTGPT: The end of online exam integrity? *arXiv* (Cornell University). https://doi.org/10.48550/arxiv.2212.09292

Thorp, H. H. (2023). ChatGPT is fun, but not an author. *Science, 379*(6630), 313. https://doi.org/10.1126/science.adg7879

Verma, V., Fleisig, E., Tomlin, N., & Klein, D. (2023, May 24). *Ghostbuster: Detecting text ghostwritten by large language models*. arXiv.org. https://arxiv.org/abs/2305.15047

Waltzer, T., Cox, R. L., & Heyman, G. D. (2023). Testing the ability of teachers and students to differentiate between essays generated by ChatGPT and high school students. *Human Behavior and Emerging Technologies, 2023*, 1–9. https://doi.org/10.1155/2023/1923981

Young, J. R. (2024, July 2). As more AI tools emerge in education, so does concern among teachers about being replaced. *EdSurge*. https://www.edsurge.com/news/2024-07-02-as-more-ai-tools-emerge-in-education-so-does-concern-among-teachers-about-being-replaced

Part III
Academic Integrity Policies

Chapter 9
Developing Academic Integrity-Compliant Regulations and Policies on the Use of Generative AI in Higher Education: Insights from the United Kingdom

Dimitar Angelov

Abstract By exploring how UK universities, and the country's higher-education sector at large, have responded to the widespread usage of ChatGPT and other large language models, this chapter will highlight the ethical and integrity challenges that higher education institutions need to bear in mind when seeking to regulate AI-informed teaching and research. Specifically, the chapter will adopt an academic-writing and integrity perspective to highlight aspects of existing AI guidelines which give rise to questions and concerns. As the UK seeks to position itself as a rule setter in the development and implementation of AI technology, and as the British education sector has embraced the use of AI-assisted tools with cautious optimism, the insights from UK national and institutional debates can contribute to a global understanding of the challenges facing sector-wide policy alignment and compliance.

Keywords Academic integrity · Ethics · Artificial intelligence · Higher-education policy · Academic writing

Introduction

With the prospect of increased regulatory divergence from the European Union (EU) post-Brexit, and with its strategic position between Europe and North America, not only geographically, but also in terms of political, cultural and economic alignment, the United Kingdom (UK) is seeking to claim a stake in the development of artificial intelligence (AI) as the next frontier of information-technology (IT) innovation. Since the launch of ChatGPT in November 2022, the UK government has published an AI strategy document (Department for Science, Innovation &

D. Angelov (✉)
Research Centre for Global Learning, Coventry University, Coventry, UK
e-mail: dimitar.angelov@coventry.ac.uk

A. E. King (ed.), *Artificial Intelligence, Pedagogy and Academic Integrity*, Advances in Artificial Intelligence in Education 1,
https://doi.org/10.1007/978-3-031-92534-4_9

Technology, 2023a) and hosted a high-profile international summit on AI safety, 1–2 November 2023, (UK Government, n.d.) in an effort to position the country as a global AI hub and a trendsetter in the development of the new technology. The government strategy sent a strong signal to the corporate sector that the UK is ready to embrace innovations with all the potential benefits they can bring (Department for Science, Innovation & Technology, 2023a, pp. 4–5). This strong pro-business ethos was also in evidence at the November AI Safety Summit, which followed. The summit brought together the major political and corporate players in the global IT industry—countries such as the United States (US) and China, alongside the EU, and sector leaders such as Google, Meta and Microsoft, to mention only a few—with the aim of agreeing on or at least debating the possibility of a set of rules that can regulate, and thereby constrain, the future development of AI-assisted technology (UK Government, 2023a). Domestically, the meeting was hailed as a success as it opened up a dialogue on sector-defining issues amongst competing global interests in AI development (UK Government, 2023a). It enabled the signing of the Bletchley Declaration, named after the venue of the summit, which was hailed as "a world-first agreement" endorsed by 28 states "establishing a shared understanding of the opportunities and risks posed by frontier AI and the need for governments to work together to meet the most significant challenges" (UK Government, 2023a). The summit defined "frontier AI" as "highly capable general-purpose AI models that can perform a wide variety of tasks and match or exceed the capabilities present in today's most advanced models" (Department for Science, Innovation & Technology, 2023b). Included in this definition were large language models (LLMs) such as ChatGPT, Claude and Bard (Department for Science, Innovation & Technology, 2023b). In addition to the Bletchley Declaration, the forum launched a UK-based AI Safety Institute (UK Government, 2023b) which, alongside a US counterpart (The White House, 2023), has the ambition to shape the growth and development of the global AI industry.

The AI Safety Summit can be seen as emblematic of the direction in which UK's national and sector AI strategies will evolve which, in turn, will impact on policies developed by higher-education regulators and providers. Despite its apparent success, the summit has raised questions about the respective roles that the private and public sectors will play in the development and regulation of AI technology, which will be all-pervasive and will have the power to shape every aspect of human activity, including research and education (Narayanan, 2023). A comment by Demis Hassabis, co-founder and CEO of Google DeepMind, (Google DeepMind, n.d.) has captured the gravity of the choices lying ahead: "[…] 'move fast and break things' has worked in building massive tech companies providing great services, but AI is too important a technology to do it in that way" (Tuck, 2023). With such high stakes in play, Hassabis's words are an urgent call for action to a wide range of stakeholders who need to be involved in overseeing the future applications of AI. It is also a warning of the possible threat that the IT industry can pose to sectors, such as education, which move at a different pace and comply with different professional and ethical standards. The enthusiasm and urgency demonstrated at the AI Safety Summit and the speed with which AI innovations are already being incorporated into existing occupations and IT infrastructure will make it progressively harder for UK universities to safeguard against any adverse effects of the technology, with all the ensuing risks to the integrity of their assessment and degrees.

The UK Higher Education Landscape

The launch of ChatGPT in November 2022 sent shockwaves across UK higher education, with local universities scrambling to ensure the validity of their assessment practices. One amongst the first coherent responses in the sector came from the UK Quality Assurance Agency for Higher Education (QAA), an independent expert body for tertiary education (QAA, n.d.), which published a set of guidelines aimed at higher education providers in January 2023 (QAA, 2023a). This early document framed the use of generative AI, LLMs and specifically ChatGPT, as primarily a threat to university teaching and learning, and set out to suggest countermeasures. For the QAA, the main problem that this new technology posed was one of textual authorship; the fact that the output of LLMs was unique and could not be identified by 'more traditional plagiarism detection tools' meant that students could submit for assessment content generated by AI, instead of their own work (QAA, 2023a, p. 1). To address this issue, the Agency offered a set of containment and mitigation strategies which included: raising awareness of the unreliability of the technology, as well as of the long-term damage cheating does to students' education; promoting declarations verifying the authenticity of students' assignments; reviewing assessment design to minimise opportunity for submitting AI-generated text; and updating existing institutional policies to prohibit such submissions (QAA, 2023a, p. 1). Notwithstanding the sensible and pedagogically sound nature of these measures, however, they did not offer a definitive solution to students' misuse of generative-AI tools, since, according to the QAA's own admission, there was no reliable protocol that could be used to ascertain textual authorship.

The overall tone of this early document is thus one of caution, if not consternation, at the latest technological developments, and, although it indicates that generative AI has "potential benefits […] for learning, teaching and assessment" (QAA, 2023a, p. 1), it focuses on what seems like damage limitation rather than on future opportunities. As such, it is an accurate representation of the tense atmosphere that seemed to prevail in the UK higher-education sector at the time. Soon after the launch of ChatGPT, the majority of UK universities rushed to issue bans on the technology in response to public panic over the integrity of academic assessments and degrees. In early 2023, the prevalent advice was for students *not* to use generative AI in their assignments (Webb, 2023). An article in the popular British daily *The i*, published on 28 February 2023, estimated that nearly 40% of UK universities have either prohibited the use of ChatGPT in student assessment or were in the process of doing so (Wood, 2023). Amongst these were the most elite universities in the country, Oxford and Cambridge, as well as other high-ranking institutions such as Manchester, Bristol and Edinburgh. The article also mentioned outliers, however, such as the University of Glasgow and University College London (UCL), also top-ranking institutions (Times Higher Education, 2024), which took a more balanced approach and chose to allow a limited use of the technology. Glasgow academics were reportedly interested in "how students can be advised to use large language models responsibly in their coursework," whilst the UCL website was referenced as giving concrete examples of how ChatGPT could be employed by students, namely, "to save time on usually laborious bibliographies, acknowledgements and references" (Wood, 2023). The editing capabilities of ChatGPT received another positive, albeit cursory, mention on the website of Jisc, "the UK digital, data and technology agency focused on tertiary education, research and innovation" (Jisc, n.d.), which advocated the use of the tool in a limited number of cases, but without going into detail (Webb, 2023).

Once the initial impact of ChatGPT's launch started to wane, the academic consensus in the UK began to shift towards a more open-minded view of generative AI, and, consequently, a more in-depth discussion of its pedagogic benefits ensued. In May 2023, the QAA issued a second briefing paper which sought to update and build on its earlier guidelines. The Agency acknowledged that several factors had changed since the January document: "not only has the technology evolved, but there has been considerable reflection and debate in the sector around how these tools might be used to support academic programmes while maintaining the integrity of awards" (QAA, 2023b). Last, but not least, was the admission that the all-pervasiveness of generative AI, combined with the lack of detection capability, had rendered any attempt to prohibit its use futile (QAA, 2023b, p. 1).

The QAA May briefing note distinguishes between appropriate and inappropriate use of generative AI, and promotes the development of artificial-intelligence literacy which can help students to navigate university regulations and use the technology responsibly (QAA, 2023b, p. 5). It goes on to give examples of good practice such as "producing early drafts; conducting rudimentary (for now) literature searches; transcribing and summarising lectures in a student's first language; and helping students prepare final versions of their submissions" (QAA, 2023b, p. 5). Additionally, the QAA enumerates instances where universities are already dealing with what it calls "hybrid submissions" in which students have used generative AI as "an assistive technology, to generate initial ideas, or to refine the final submission by correcting grammar/spelling, or removing redundant text to meet a word limit" (QAA, 2023b, p. 2). Yet the document remains vague on what constitutes illegitimate uses of AI. The only example it provides is a rather extreme case of student assignments that "consist only of substantially unmodified output from Generative Artificial Intelligence software", but even this infringement of what would normally be understood as authorship is described tentatively as "very poor academic practice", rather than cheating, and the recommended course of action is additional student support instead of disciplinary measures (QAA, 2023b, p. 3). Despite introducing a distinction between appropriate and inappropriate uses of generative AI, the document steers clear of formulating any pedagogical principles that can be applied to draw a line between the two. The responsibility for any such qualifying action is passed on to UK universities instead:

> [P]roviders should have consistent and effective policies that explain how and in what circumstances Generative Artificial Intelligence tools can be used as an assistive technology to support the preparation of student submissions. They should be explicit over how that assistance is acknowledged, what constitutes acceptable academic practice, and take account of the integration of Generative Artificial Intelligence tools into software that is licensed by the provider for student use (QAA, 2023b, p. 2).

Also in May 2023, 6 months after the launch of ChatGPT, Jisc published a report, *Generative AI—a primer*, in which it outlined three strategic approaches as to how to deal effectively with the ubiquity of AI-tools in UK higher education. The first two of these, "avoid" and "outrun," were premised on the assumption that the technology should be excluded from the assessment process, while the third, "embrace and adapt," encouraged universities to make the most of what AI can offer (Jisc, 2023a, p. 9). Specifically, the "avoid" approach recommended onsite exams, during which access to AI could be restricted; "outrun" meant designing assignments that were impervious to the technology; and, as the name itself suggests, the "embrace and adapt" approach advocated that "appropriate use of AI" should become essential to assessment design. Although the benefits of each of these three approaches were counterbalanced by challenges, the third one was promoted as the soundest, not least because it had been favoured more widely in the sector (Jisc, 2023a, p. 8).

Similarly to the QAA briefing paper, published at the same time, the Jisc document demonstrates the attitude shift in UK higher education towards acceptance of and engagement with the new technology. In support of its "embrace and adapt" ethos, and very much like the QAA May document, Jisc provides examples of acceptable pedagogical uses of AI which are compliant with its understanding of teaching and learning ethics. These include: (1) "formula[ting] ideas" or "producing outlines as a starting point for an assignment"; (2) using AI "as a research tool" with the proviso that the technology has a "tendency […] to give misinformation"; (3) generating "feedback on style and content"; and (4) creating images which can be incorporated into students' coursework (Jisc, 2023a, p. 10).

By presenting a range of possibilities for the use of generative AI but without stipulating guiding principles for decision-making, Jisc effectively joined the QAA in leaving the ultimate say of what constitutes ethical and academic integrity misuse of the technology to individual universities. Although understandable in view of the mostly advisory role of both organisations, such a hesitant approach creates potential problems for both providers and the country's higher education sector as a whole. Universities are left to negotiate a very complex and fast-changing landscape in which different pedagogical and professional practices, moral codes, and regulations intersect. To compile a comprehensive policy on generative AI use, an institution must have recourse to advanced expertise in AI design and implementation, ethics and integrity, as well as academic writing pedagogy, and at the same time be able to mobilise and coordinate this expertise successfully. Given that in UK institutions this broad range of knowledge is split across various units with their own agendas and priorities that do not necessarily align (e.g., academic faculties and departments, academic development teams, libraries and skills units, student support and other services), a concerted response to the global challenge of AI would be extremely time- and resource-intensive. Furthermore, it may lead to diverging practices across the sector that can undermine the compatibility of awards.

To ensure consistency of approach, the top 24 research-led universities in the UK, known collectively as "the Russell Group" (Russell Group, n.d.) issued a statement in July 2023 in which they endorsed "the ethical and responsible use of generative AI," and formulated five broad principles as guidance:

1. Universities will support students and staff to become AI-literate.
2. Staff should be equipped to support students to use generative AI tools effectively and appropriately in their learning experience.
3. Universities will adapt teaching and assessment to incorporate the ethical use of generative AI and support equal access.
4. Universities will ensure academic rigour and integrity is upheld.
5. Universities will work collaboratively to share best practice as the technology and its application in education evolves [*sic*] (Russell Group, 2023, p. 1).

Although undoubtedly a step in the right direction, the document offers little in terms of concrete rules or actions that can model best practice of AI use in the classroom. It holds out a promise that universities will develop "institution-wide policies" but relegates the responsibility for their development to individual providers (Russell Group, 2023, p. 2). Such a decision seems to run counter to the *raison d'être* of the document itself, which was to provide a coordinated response to the use of AI in UK higher education in the first place.

Furthermore, the statement leaves the implementation of its recommended "institution-wide policies" to academic departments, and even individual lecturers, citing as reasons a range of context-specific factors that can influence on-the-ground pedagogy (e.g., disciplinary differences; student demographic groups and their learning needs; and non-academic stake holders, such as professional bodies, whose standards inform the teaching and learning on specific programmes) (Russell Group, 2023, p. 2). Although these concerns are indeed legitimate, they should not deflect attention from the paucity of detail and practical solutions on offer in the document. True, it is ultimately front-line university staff such as disciplinary lecturers, student-support workers and administrators who will make decisions on what constitutes appropriate or inappropriate use of AI in assessment, but such practical circumstances should not absolve advisors and senior administrators from the responsibility to provide clear and specific guidance, including examples, to be used as points of reference. Otherwise, the room for subjectivity and inconsistency in safeguarding the rules will increase exponentially.

As this brief overview has shown, actionable steps and substantiated guiding rules are conspicuously few and far between in all of the aforementioned documents that advocate responsible use of AI in higher education. When concrete suggestions are made, they often appear inconsistent and even contradictory. At the heart of this problem seems to be the lack of precise definitions as to what constitutes "responsible use" from the point of view of ethics and academic integrity, with the issue of "authorship" providing a perfect illustration in this regard.

Authorship as an Integrity Litmus Test

As discussed above, the QAA's only clear example of academic-integrity rules infringement is a case where "substantially unmodified output" from generative AI is submitted for assessment. In accordance with this view, the intellectual ownership of a text is determined solely by whether or not its formal consistency and integrity as a finished artefact can be truthfully attributed to its designated author(s). This premise rests on the assumption that a text can be reduced to the end product of writing, rather than being a complex semantic structure that evolves through various stages of authorial intervention. In the scholarship of academic writing, the latter view has been explored at length as part of the so-called "process pedagogy" which frames writing as a process of discovery whereby meaning is constructed through several iterative stages, rather than appearing ready-made as the product of mechanically put-together words on a page (Anson, 2014). A pioneering scholar of the process pedagogy, Donald Murray, conceived of writing as, variously, consisting of three distinct stages: "prewriting", "writing", and "rewriting" (Murray, 1972/2009, p. 2), or four such stages: "invention", "drafting", "revising" and "editing" (Murray, 1984, as cited in Anson, 2014, p. 223). The proliferation of taxonomies and approaches to writing as a process in the 1970s and 1980s, in particular, gave rise to the so-called "process movement," which remained the dominant model of teaching writing at university level for decades, and is still influential, if not leading, in classrooms across Anglophone academia and further afield (Hyland, 2016, pp. 12–18).

A detailed discussion of the process movement is, of course, beyond the scope of the present study; however, even a brief overview of the recursive stages of writing it proposes highlights potential problems with the existing advice on the responsible and ethical use of generative AI for assessment and research. Amongst the QAA's examples of the technology being used appropriately were carrying out literature searches, generating ideas for writing, composing early drafts, adjusting the length of the text to meet assessment word-limit requirements, as well as improving the quality of the finished product in terms of grammar and spelling. Although the last example of textual intervention has been an automated feature of most word-processing programs for years, all of the other procedures have been made possible by generative AI and could be ascribed to one or more of the stages of the writing process. An automated mining of the literature by an AI-assisted tool and a concurrent generating of ideas for writing have been traditionally seen as "prewriting" tasks, part of the "invention" or "planning" of the textual content. Producing early drafts of a text will involve "planning" and will, of course, come under "drafting", whilst resizing a text is a classic example of "revising", which usually requires strategies such as summarising and paraphrasing. Similarly, the uses recommended by Jisc, namely, to research a particular topic, generate ideas and outlines for writing, and provide feedback on drafts can be clearly classified as "inventing", "planning" and "drafting" tasks.

Both QAA and Jisc seem to view these steps as purely mechanical operations which can be optimised with the help of technology, rather than as complex cognitive processes which require human agency and decision-making. Writing, in all of its stages, is not simply an automated sequence in which information is sorted and matched according to the probability rules of language models, but a process where ideas develop through the critical thinking and evaluation of a socially-situated and embodied human intelligence. To renounce authorial control over the *entirety* of the writing process by delegating tasks to an external agent is to compromise the concept of authorship as it is currently understood. As the process of writing, in all its complexity, is in fact a process of authoring the text.

In support of this stance are the guidelines on authorship and the use of AI issued by some of the leading international publishers such as Elsevier and the Science journal group. The current Elsevier publishing policy explicitly prohibits the use of generative AI in the writing process, apart from the instances where surface features of the text are concerned. Interventions in the argument of a research paper, on the other hand, are strictly off limits: "Where authors use AI and AI-assisted technologies in the writing process, these technologies should only be used to improve readability and language of the work and not to replace key authoring tasks such as producing scientific, pedagogic, or medical insights, drawing scientific conclusions, or providing clinical recommendations" (Elsevier, 2024).

In an editorial published in *Science*, H. Holden Thorp, editor-in-chief of the Science journal group goes even further in defining inappropriate practices:

> For years, authors at the *Science* family of journals have signed a license certifying that "the Work is an *original*" (italics added). For the *Science* journals, the word "original" is enough to signal that text written by ChatGPT is not acceptable: It is, after all, plagiarized from ChatGPT. Further, our authors certify that they themselves are accountable for the research in the paper. Still, to make matters explicit, we are now updating our license and Editorial Policies to specify that text generated by ChatGPT (or any other AI tools) cannot be used in the work, nor can figures, images, or graphics be the products of such tools. And an AI program cannot be an author. A violation of these policies will constitute scientific misconduct no different from altered images or plagiarism of existing works (Thorp, 2023, p. 313).

The current *Science* journals' policy, as found on the publisher's website, stops short of an outright ban on the use of generative AI for text production, but imposes conditions that could be difficult for authors to meet. For example, authors are responsible for ensuring not only the accuracy of the output, but also that it contains no instances of plagiarism, and that it is appropriately referenced and unbiased (*Science*, 2024). A similar position is also taken by Taylor & Francis, which requires authors to guarantee "the originality, validity and integrity of the content of their submissions," including any content produced by AI-assisted tools (Taylor & Francis, 2023). Given that generative AI tools have been trained on vast corpora of publicly available data with often unclear provenance and that they operate on the so-called "black-box" principle whereby their internal processes are unpredictable and untraceable (Lanier, 2023), any declaration of clarity regarding ownership of ideas in AI-generated outputs will be unreliable, at best, or outright misleading.

Publishers' guidance on AI-produced textual content can paradoxically contradict their recommendations regarding AI-generated images. At the time of writing of this chapter, Elsevier (2024), Science (2024), and Springer (2023) allow at least some use of the technology for textual-content creation, whilst strictly prohibiting it for visual content. The logic as to why the production of images is viewed as being different from the production of text remains unclear and unjustified. One could speculate that a copyrighted image is easier to identify than a verbal argument, which could be transformed through re-ordering or paraphrasing, making integrity breaches of visual material, and hence legal liability, easier to prove. In any case, instead of a distinction made on principle, taking into account agency and creativity as the drivers of both visual and verbal content creation, the rules of AI use here seem purely arbitrary.

There is one rule over which most publishers have reached a formal consensus and it is that ChatGPT, alongside similar text-generating chatbots, cannot be referenced as an author—see Elsevier (2024), Science (2024), Springer (2023) and Taylor and Francis (2023). However, given that the advent of AI tools is already disrupting the idea and practice of autonomous authorship, the utility of this recommendation, beyond a purely formal gesture of attribution, remains doubtful.

Further Ethical Considerations

Authorship aside, the inclusion of AI-assisted tools in academic practice poses a host of other challenges related to ethics and integrity. Mentioned almost in passing in some of the earlier UK guidelines published in 2023, these challenges received greater attention as the higher-education-sector views shifted towards a gradual acceptance of generative AI in teaching and research. If the first QAA briefing note published in January 2023, focuses exclusively on the danger of students passing off AI-produced text as their own (QAA, 2023a), the Agency's May document refers to potential breaches of data protection as an additional risk to consider, and elaborates on new equity issues arising from the introduction of generative AI into UK universities (QAA, 2023b). Since most AI-assisted tools have a more basic (free) and a more advanced (paid-subscription) version, the QAA predicted that students from wealthier socio-economic backgrounds would gain an unfair advantage by being able to use the latest and most powerful technology, whilst most working-class and ethnic-minority students would risk falling behind even further (QAA, 2023b, p.5).

Similarly, since May 2023, Jisc has increasingly raised awareness of ethical and integrity challenges, apart from ascertaining authorship, that generative AI poses for higher-education institutions. It first presented only a brief list of limitations: biases inherent in the data sets used by the technology; lack of transparency regarding the sources and ownership of information; and negative impact on the environment and people, including disadvantages caused by digital inequality (Jisc, 2023a, p. 6). Then these were articulated more clearly, and in greater detail, in the subsequent Jisc policy document, published in September 2023 (Jisc, 2023b, pp. 5–6).

The trend was continued in the Russell Group guidelines for AI use, also published in September 2023 (Russell Group, 2023, pp. 1–2). Apart from clarifying further the human cost of the development and deployment of AI-assisted tools (for example, by mentioning the harmful working conditions which operators may experience whilst policing AI-training data sets against extreme material), the document provides an important overarching warning: "[U]sers of generative AI tools should be aware that while ethics codes exist, they may not be embedded within all generative AI tools and that their incorporation, or otherwise, may not be something that users can easily verify" (Russell Group, 2023, p. 2). The quotation makes an important admission which most AI-policy writers in UK higher education tend to omit, namely, that by incorporating AI technology into academic work users implicitly accept the ethical and integrity standards of the tools they access. In other words, the ethics of the global IT industry, and specifically the companies which produce AI-assisted tools, has the potential to dominate and, at times, undermine any national or institutional policies that UK higher-education sector has or may adopt in the future. Short of developing generative AI tools that have been designed specifically for UK universities, any other software product may introduce principles and norms that belong to different national or sector regulatory regimes. Such a state of affairs creates vulnerabilities for individual users, in particular, such as students, researchers, university lecturers and support staff, because it is the end-users who are entrusted with the ultimate responsibility for how to deploy any AI-assisted tool.

Given the complicated and opaque nature of the technology, the speed with which it has been introduced into UK higher education and the multiple economic interests it represents and impacts on, it is unclear to what extent individual users can make any reliable judgments of its potential risks, or make valid claims about its ethical use. Cascading the responsibility for decision-making and action to higher-education frontline staff and students can thus be seen as an attempt to avoid uncomfortable truths, or even liability.

Conclusion

The review of guidelines offered in this chapter aims to provide a representative account of the ethical and integrity challenges that the UK higher-education sector has faced since the 2022 launch of ChatGPT and the rapid introduction of generative AI in university teaching and research practices. UK higher education can be seen as an informative case study for both the potential and the challenges of the new technology due to the government's low-regulation and pro-business approach, which is likely to incentivise the technology's more rapid and pervasive deployment. The UK's decision not to pre-empt any use of generative AI through early restrictions and to impose stricter rules only when a need has been proven can be seen reflected in the ethos of its higher-education institutions. After an initial moment of consternation, UK universities have become open to and even embraced the new technology, seeking to make the most of a situation which is effectively beyond their control.

However, whilst modifying existing academic-integrity guidelines and regulations to accommodate generative-AI use, UK higher-education agencies and providers have raised further questions and left many issues partially or fully unaddressed. Concepts like authorship and originality in writing and image production have become destabilised with the advent of AI and are in need of urgent re-consideration and re-defining. Similarly, ethical problems embedded in the design of the available software products have been highlighted but remain unresolved, and therefore continue to undermine any efforts on the part of end-users to conform to accepted academic standards. The public advice of UK higher-education policy-makers and providers leaves the overwhelming sense that the sector is ill at ease with the new technology and that its efforts to adapt have been plagued by internal inconsistencies and contradictions. Developed internationally and in compliance with different sets of industry standards and rules, AI-assisted tools will always have the potential to breach the codes of conduct of a nation state's education sector. To remedy regulatory misalignments of this kind will require good will and dialogue amongst a number of stakeholders from different political and economic contexts, which seems difficult to achieve at this stage. For the time being, universities in the UK and elsewhere will have to follow the lead of Big-Tech decision-makers, with the option to either only partially mitigate the ethical and integrity risks generative AI poses for them, or reinvent their existing notions of academic ethics and integrity altogether.

References

Anson, C. M. (2014). Process pedagogy and its legacy. In G. Tate, A. Rupiper, K. Schick, & H. B. Hessler (Eds.), *A guide to composition pedagogies* (2nd ed., pp. 212–230). Oxford University Press.

Department for Science, Innovation & Technology. (2023a, March 29). *A pro-innovation approach to AI regulation*. Retrieved July 5, 2024, from https://assets.publishing.service.gov.uk/media/64cb71a547915a00142a91c4/a-pro-innovation-approach-to-ai-regulation-amended-web-ready.pdf

Department for Science, Innovation & Technology. (2023b, October 25). *Frontier AI: Capabilities and risks – Discussion paper*. Retrieved July 5, 2024, from https://www.gov.uk/government/publications/frontier-ai-capabilities-and-risks-discussion-paper/frontier-ai-capabilities-and-risks-discussion-paper

Elsevier. (2024). *The use of AI and AI-assisted technologies in writing for Elsevier: Policy for journal article, book and commissioned content authors*. Retrieved July 9, 2024, from https://www.elsevier.com/about/policies-and-standards/the-use-of-generative-ai-and-ai-assisted-technologies-in-writing-for-elsevier

Google DeepMind. (n.d.). *Our mission: Build AI responsibly to benefit humanity*. Retrieved July 5, 2024, from https://deepmind.google/about/?_gl=1*xjutwi*_up*MQ..*_ga*NjgwNzQ1MjQwLjE3MDU0NDUxNTg.*_ga_LS8HVHCNQ0*MTcwNTQ0NTE1Ny4xLjAuMTcwNTQ0NTE1Ny4wLjAuMA

Hyland, K. (2016). *Teaching and researching writing* (3rd ed.). Routledge.

Jisc. (2023a, July 12). *Generative AI – a primer*. Retrieved July 5, 2024, from https://www.jisc.ac.uk/reports/generative-ai-a-primer

Jisc. (2023b, September 6). *AI in tertiary education* (3rd ed.). Retrieved July 5, 2024, from https://www.jisc.ac.uk/reports/artificial-intelligence-in-tertiary-education

Jisc. (n.d.). *About us*. Retrieved July 5, 2024, from https://www.jisc.ac.uk/about-us

Lanier, J. (2023, April 20). There is no AI. *The New Yorker*. https://www.newyorker.com/science/annals-of-artificial-intelligence/there-is-no-ai

Murray, D. (2009). Teach writing as a process not product. In T. Newkirk & L. C. Miller (Eds.), *The essential Don Murray: Lessons from America's greatest writing teacher* (pp. 1–5). Boyton/Cook Publishers. (Reprinted from *Leaflet*, pp. 11–14, 1972, November).

Narayanan, M. (2023, October 23). Transparency seems to die a death at the government's AI summit. *Byline Times*. https://bylinetimes.com/2023/10/23/transparency-seems-to-die-a-death-at-the-governments-ai-summit/

Quality Assurance Agency for Higher Education. (2023a, January 30). *The rise of artificial intelligence software and potential risks for academic integrity: A QAA briefing paper for higher education providers*. Retrieved July 5, 2024, from https://www.qaa.ac.uk/docs/qaa/members/the-rise-of-artificial-intelligence-software-and-potential-risks-for-academic-integrity.pdf?sfvrsn=ebb0a981_6

Quality Assurance Agency for Higher Education. (2023b, May 8). *Maintaining quality and standards in the ChatGPT era: QAA advice on the opportunities and challenges posed by generative artificial intelligence*. Retrieved July 5, 2024, from https://www.qaa.ac.uk//en/membership/membership-areas-of-work/generative-artificial-intelligence/qaa-advice-and-resources

Quality Assurance Agency for Higher Education. (n.d.). *The UK's expert quality body for tertiary education*. Retrieved July 5, 2024, from https://www.qaa.ac.uk

Russell Group. (2023, July 4). *Russell Group principles on the use of generative AI tools in education*. Retrieved July 5, 2024, from https://russellgroup.ac.uk/news/new-principles-on-use-of-ai-in-education/

Russell Group. (n.d.). *Who we are*. Retrieved July 5, 2024, from https://russellgroup.ac.uk/about/

Science. (2024). *Science Journals: Editorial Policies*. Retrieved July 9, 2024, from https://www.science.org/content/page/science-journals-editorial-policies#image-and-text-integrity

Springer. (2023). *Artificial Intelligence (AI)*. Retrieved July 9, 2024, from https://www.springer.com/gp/editorial-policies/artificial-intelligence%2D%2Dai-/25428500#:~:text=While%20legal%20issues%20relating%20to,in%20a%20legally%20acceptable%20manner

Taylor & Francis. (2023). *Taylor & Francis clarifies the responsible use of AI tools in academic content creation*. Retrieved July 9, 2024, from https://newsroom.taylorandfrancisgroup.com/taylor-francis-clarifies-the-responsible-use-of-ai-tools-in-academic-content-creation/

The White House. (2023, November 1). *Remarks by Vice President Harris on the future of artificial intelligence*. Retrieved July 5, 2024, from https://www.whitehouse.gov/briefing-room/speeches-remarks/2023/11/01/remarks-by-vice-president-harris-on-the-future-of-artificial-intelligence-london-united-kingdom/

Thorp, H. H. (2023). ChatGPT is fun, but not an author. *Science, 379*(6630), 313. https://www.science.org/doi/10.1126/science.adg7879

Times Higher Education. (2024, February 23). Best universities in the UK 2024 – University rankings. https://www.timeshighereducation.com/student/best-universities/best-universities-uk

Tuck, D. (2023, December 8). 'Move fast and break things' doesn't apply to AI. *Forbes*. https://www.forbes.com/sites/forbestechcouncil/2023/12/08/move-fast-and-break-things-doesnt-apply-to-ai/?sh=446f1cd319af

United Kingdom Government. (2023a, November 1). *Countries agree to safe and responsible development of frontier AI in landmark Bletchley Declaration*. Retrieved July 5, 2024, from https://www.gov.uk/government/news/countries-agree-to-safe-and-responsible-development-of-frontier-ai-in-landmark-bletchley-declaration#:~:text=The%20Bletchley%20Declaration%20on%20AI,deployed%20in%20a%20safe%2C%20responsible

United Kingdom Government. (2023b, November 2). *Prime Minister launches new AI Safety Institute*. Retrieved July 5, 2024, from https://www.gov.uk/government/news/prime-minister-launches-new-ai-safety-institute#:~:text=The%20Institute%20will%20carefully%20test,humanity%20losing%20control%20of%20AI

United Kingdom Government. (n.d.). About the AI safety summit 2023. Retrieved July 5, 2024, from https://www.gov.uk/government/topical-events/ai-safety-summit-2023/about

Webb, M. (2023, February 14). *Considerations on wording when creating advice or policy on AI use*. Jisc. Retrieved July 9, 2024, from https://nationalcentreforai.jiscinvolve.org/wp/2023/02/14/considerations-on-wording-ai-advice/#comments

Wood, P. (2023, March 2). Oxford and Cambridge ban ChatGPT over plagiarism fears but other universities choose to embrace AI bot. *The i Paper*. https://inews.co.uk/news/oxford-cambridge-ban-chatgpt-plagiarism-universities-2178391

Chapter 10
Frozen in Time: Croatian Policies on Academic Integrity and GenAI in Higher Education

Pegi Pavletić

Abstract This chapter analyses academic integrity policies and practices at nine public universities in Croatia. Generative AI (GenAI) is in use in Croatian higher education, with the number of research projects related to artificial intelligence growing across different fields. Despite the term academic integrity not being used widely in Croatia, the Codes of Ethics and Disciplinary Codes of the universities address academic integrity, misconduct, and ethical values concerning faculty, employees and students. The findings conclude that all universities need to update their disciplinary procedures and reflect on the opportunities and threats of GenAI, as none of them directly link the use of advanced digital tools with academic misconduct. Universities must develop practices that teach students and staff members about GenAI and a culture of academic integrity.

Keywords Academic integrity · Higher education · Croatia · Artificial intelligence · Students · Disciplinary procedures

Introduction

Generative Artificial Intelligence (GenAI) is able to generate content, media or other outputs (e.g., codes) based on the way it is programmed (Eke, 2023). This means that generative AI software is able to produce content much faster than a human could, without much effort needed for research or preparation of a particular topic. It is understandable how such tools are able to gain broad acceptance and

P. Pavletić (✉)
The European Students' Union (ESU), Brussels, Belgium
e-mail: pegi.pavletic@esu-online.org

A. E. King (ed.), *Artificial Intelligence, Pedagogy and Academic Integrity*, Advances in Artificial Intelligence in Education 1,
https://doi.org/10.1007/978-3-031-92534-4_10

popularity for use in higher education; however, their use inevitably leads to concerns related to academic integrity (Eke, 2023; Foltynek et al., 2023). One of the tools that has recently gained popularity is ChatGPT, a large language model that is able to imitate human-generated responses based on deep learning (Eke, 2023). Its use in academia is controversial because the policymakers fear that this tool could be used by students to cheat; however, there are considerable debates regarding the definition of academic misconduct for the improper use of AI.

The European Network for Academic Integrity (ENAI) made the point that not all use of GenAI is to be considered unethical, as this can depend on multiple factors, such as a discipline of studies, culture, national systems, and so on (Foltynek et al., 2023). They argue that GenAI tools need to be acknowledged when they are used in content/output generation. Furthermore, they highlight the importance for Higher Education Institutions (HEIs) to develop their policies on the use of GenAI (potentially based on national recommendations), as well as provide training to teachers, students and other learners on the correct use of GenAI.

When it comes to the use of GenAI in Higher Education (HE), some Croatian universities have been early adopters. The need for the introduction of GenAI in medical faculties was demonstrated by the implementation of individual courses on Medical Informatics at the universities in Zagreb, Rijeka, Osijek and Split in 2009 (Kern & Petrovečki, 2009). In 2021, the students' use of ChatGPT3 at the University Department of Forensic Sciences in Split was evaluated by Bašić et al. (2023a), and it was found that those students who did not rely on ChatGPT3 for their academic assignments had only a slightly better score than those who used the tool. However, the students using ChatGPT3 exhibited more issues with plagiarism (Bašić et al., 2023a). Due to the lack of consistency in national practices of using GenAI, I assess regulations and policies connected with GenAI use at nine public Croatian universities. Additionally, I evaluate whether links were made to academic integrity concerns related to the use of GenAI and if students were involved or acquainted with these regulations and policies.

Research Objectives

In Croatia, the use of GenAI expanded rapidly after the introduction of ChatGPT (Milinković et al., 2024; Bašić et al., 2023b); I explore whether the Croatian public universities recognize the need for monitoring, regulating and/or educating their staff and students on the use of GenAI and its connection to the values of academic integrity. This was particularly relevant as higher education is training highly skilled individuals who will have the potential to impact their relative working fields, but also collaborate with their younger peers on activities which could involve the use of GenAI tools. Without policies in place guiding the use of GenAI tools, there is a possibility of increased reliance on GenAI by higher education students and graduates, leading to the lack of obtained work-related or research-related skills, as well as academic misconduct. Aside from the impact on students, educators may benefit from clear guidelines and policies on the use of GenAI, as they assess and evaluate students' work. Policies and regulations may also help instructors to determine what is misconduct when it comes to the use of GenAI tools by students. Given the complexity and importance of having well-planned and robust policies, both students and educators should be involved in the policymaking. Therefore, two research questions were posed:

(i) Have the academic integrity policies and strategic documents at Croatian public universities recognized the existence and the use of generative-AI and included or mentioned regulations for its use?
(ii) Do the policies of the Students' Councils mention and/or regulate students' use of GenAI tools; if they do, do students' ombudspersons have a role in sharing the knowledge of academic integrity, particularly related to the use of GenAI?

At the beginning of my research, I expected that the universities would have policies in practice on the use of GenAI, and that there would be plans for introducing them in their strategic planning. Additionally, I expected that the growing use of GenAI would attract students' interest and that they would have activities related to the topic, even if the students' policies did not reflect GenAI usage. This is relevant because the policies adopted by students have to be approved at the HEI level, which could take some time, while their activities could be more descriptive of the students' interest and engagement in the topics related to the use of artificial intelligence. Therefore, these practices could paint a clearer picture about students' involvement in the topics related to both academic integrity and the use of GenAI than the policies themselves. Because student representatives come from different backgrounds and levels of education, they might not perceive that the topic of GenAI is relevant in all study fields or at all study levels of study, and that policies are therefore unnecessary for student organizations.

Methodology

I undertook a theoretical analysis of institutional policies related to academic integrity at nine public universities in Croatia to explore whether the use of GenAI was included. In addition, I examined the policies of student unions for references to GenAI. The documents analysed included the universities' Statutes, Codes of Ethics (CoEs), Disciplinary Codes (DCs), Strategy documents, Statutes of their Students' Councils[1] and the Plans of Work of the Students' Councils (if available). In the analysis of institutional policies, only the available documents on the webpages of the universities were analysed. If no document (CoE, DCE, DCS) was published, for the purpose of this research, it was considered that such document does not exist.

[1] Students' Councils at each University have different names in English, not standardised on a national level. Therefore, for simplicity, their names have all been standardised in this article.

Universities were not contacted individually because, due to their public nature, their governance documents are expected to be public; May 31st 2024 was the cut-off date for the collection of the documents. Although there is a possibility that individual Faculties belonging to the respective HEIs implement additional practices, this analysis was out of the scope of this research. Finally, a critical reflection is provided based on the findings.

Higher Education in Croatia

Croatia is a country located in south-eastern Europe with a population of roughly 3.9 million inhabitants; it is a member of the European Union (EU) and a member of the European Higher Education Area (EHEA). Under EHEA, Croatia is a signatory of the Bologna Process, signing in 2001 (The European Higher Education Area, 1999). The Bologna Process encompasses three levels of higher education (undergraduate, graduate, postgraduate) and employs a common system of credits: European Credit Transfer and Accumulation System (ECTS). Additionally, under the Bologna Process, common goals for quality assurance in higher education are agreed upon. Croatian universities also have the possibility to join the European University Association to work towards common international goals in the field of higher education. The structure of the EHEA is broad, but there are several constituencies with constant links to national systems, as represented in Fig. 10.1.

Similarly, students have the possibility to join the European Students' Union, an umbrella organization, composed of national unions of students, that promotes student participation in EHEA. A general structure of the organization of student representatives on a European and Croatian level is given in Fig. 10.2.

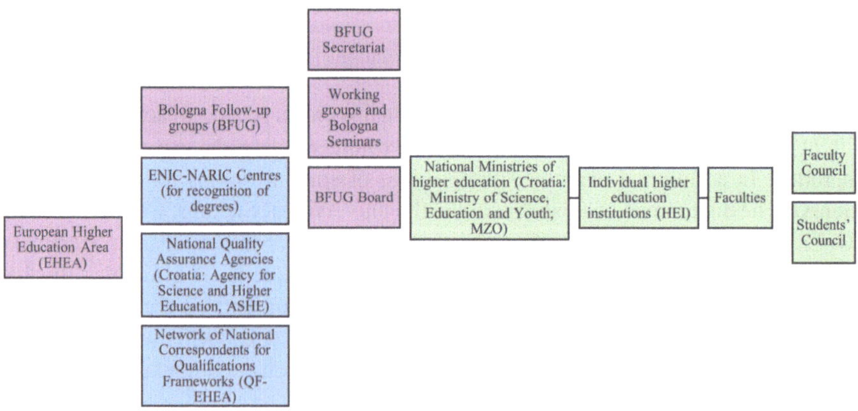

Fig. 10.1 A scheme of the structure of the European Higher Education Area constituencies (pink) and collaborators (blue). A simplified model of the Croatian higher education structure (green). The yellow dash indicates collaboration between institutions, on the matters of higher education. The red dash indicates the regulatory role of ASHE over individual higher education institutions

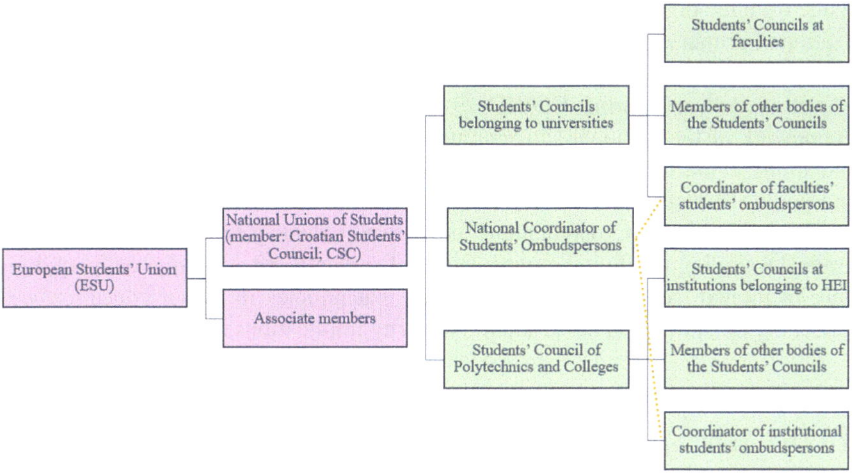

Fig. 10.2 A scheme of the structure of the student representation bodies on the European level (pink), and a model of the national students' representation structure in Croatia (green). The orange dash indicates collaboration between students' ombudspersons

To support the activities of the Bologna Process between Ministerial conferences, Bologna Follow-Up Groups were established with representatives from members of EHEA and other stakeholders. Each member nation's ministry in charge of higher education participates in the Ministerial Conferences, where further developments and commitments are discussed. Based on those commitments, Ministries, as well as other higher education stakeholders (e.g., HEIs, Quality Assurance agencies, students), propose activities in their strategic documents or plans of work.

Digital Literacy of Croatian Youth

Across Croatia, there were 151,827 students enrolled in higher education in the 2022/2023 academic year (Croatian Bureau of Statistics, 2023). According to Eurostat, Croatia had the overall highest digital literacy rate (97%) of young people (ages 16–24) in 2019 (Eurostat, 2020). Furthermore, the Croatian Digital Economy and Society Index (DESI) 2022 Report states that the overall digital literacy of Croatian population (ages 16–74) was higher than the average for the EU, with 63% of people possessing at least basic digital skills (compared to the EU average of 54%) (The European Commission, 2022). The University Computing Centre (SRCE) is the oldest (established in 1971) Croatian institution in the field of applied Information and Communication Technologies (ICT). In 2007, the SRCE established an E-learning Centre and the Merlin platform for e-learning in higher education, facilitating e-learning at more than 115 Croatian HEIs so far (Moodle, 2019).

Aside from the digitalization of higher education, efforts were being put into digitalization of primary and secondary education. For example, the e-School Programme has been launched as a pilot project for facilitating online primary and secondary education delivery during COVID-19 (The European Commission, 2022). The Framework of the National Curriculum was created to ensure that digital literacy and ability to use technology are considered to be generic competencies of children and youth (Ministry of Science and Education of the Republic of Croatia, 2016). The digital literacy of Croatian higher education students is high, with courses at all levels developing the digital competencies of students and education delivery digitally facilitated. As a result, it is expected that policies in higher education will reflect digitalization and the academic integrity implications related to it.

An Overview of Academic Integrity and Artificial Intelligence (AI) Research in Croatia

The 2017 'South East European Project on Policies for Academic Integrity' (SEEPPAI) project analysed Croatian policies for academic integrity and their effectiveness prior to the advent of GenAI (Lancaster et al., 2017). The main findings from this study illustrate that ministries and national Quality Assurance (QA) agencies do not provide institutions with strong guidance on academic integrity policies, nor was there was oversight for such policies. Furthermore, the implementation of academic integrity policies and their effectiveness were unclear. The study found that punitive measures were mostly employed to prevent misconduct and, while educational courses on academic writing were generally in place, there was a lack of information and education on academic integrity. Some universities in Croatia adopted Codes of Ethics (CoE) from Western higher education institutions (HEIs); however, their practices lacked vigilance (e.g., institutions rely on students' honesty and understanding of academic integrity) (Glendinning et al., 2017). Students believed that cheating was common, even with more than 60% of students being aware of plagiarism before attending university. Most students learned how to cite and reference during their undergraduate studies (Glendinning et al., 2017).

In Croatia, most research related to ethics has focussed on medicine (Borovečki et al., 2004; Malički et al., 2019); however, some researchers have pointed to issues related to academic integrity (Čepo, 2017). One study about academic integrity in online teaching found that students were more likely to cheat when online (Katavić et al., 2016). Katavić et al. (2016) notes that efficient academic integrity policies, better interaction between students and teachers, and frequent modification of online tasks for students could mitigate cheating in online courses. At the University of Split, Bašić et al. (2019) found that students were familiar with referencing, quoting, paraphrasing and acknowledgements of authorship, but were unfamiliar with self-plagiarism, referencing personal communication and general knowledge, and writing summaries (Bašić et al., 2019). Students generally carried negative attitudes toward plagiarism, whether or not they attended a course in which plagiarism was explained. More recently, a nation-wide 2023 report by the Ministry of Science, Education and Youth (MZO) and SRCE mapped and predicted the use of AI in technology and science (Bajić et al., 2023). From the Report, it is clear that the number of publications on AI (indexed in SCOPUS) by Croatian Scientists has almost doubled since 2012, with most publications being in the fields of Computer Science, Engineering, Materials Science and Medicine.

Academic Integrity Policies in Croatia

Existing national regulations concerning academic integrity in Croatia are largely based upon European standards. The Standards and Guidelines for Quality Assurance in the European Higher Education Area (ESG) (ENQA et al., 2015) advises that the institutional integrity of national Quality Assurance (QA) agencies is necessary and QA policies must ensure that academic integrity policies are vigilant against academic fraud. In Croatia, the Act on Quality Assurance in Science and Higher Education requires that the QA standards adhere to ESGs (Zakon o osiguravanju kvalitete u visokom obrazovanju i znanosti, 2022).

In Croatia, while the term academic integrity is not used widely, several documents at all HEIs would fall under the umbrella of academic integrity. Proper academic conduct is described under the CoE principles at HEIs, while the punitive measures for academic misconduct are prescribed through the Disciplinary Codes (DCs). According to the Act on Higher Education and Scientific Activity, the "ethical behaviour of scientists, artists and teachers" (Zakon o Visokom Obrazovanju i Znanstvenoj Djelatnosti, 2022) is required: all members of the academic community must adhere to ethical principles in their work; all HEIs and science institutes must have an ethics committee and a CoE voted upon by the executive body (e.g. Senate, faculty board etc.); and, ethics committees must follow the guidelines of the National Council for Science, Higher Education and Technology Development (NVZVOTR) for minimal ethics principles (not yet established at the time of this research). A Board for Ethics in Science and Higher Education was established under MZO; this Board created a CoE for use in higher education (Committee for Ethics in Science and Higher Education, 2006), although every HEI can broaden its content for their own use.

The 2007 Law on Students' Council and Other Students' Organisations defines the role of student ombudspersons in HE (Zakon o Studentskom Zboru i Drugim Studentskim Organizacijama, 2007). Student ombudspersons are students who are appointed by their respective student councils at each HEI for a one-year mandate that can be repeated once. Each ombudsperson collects student opinions and complaints regarding their rights and communicates them to HEI bodies (Disciplinary Committee, Ethics Committee or the Council of Honour), provides counsel to students, defends students in disciplinary procedures and other related activities as defined by the HEI Statute. Each HEI must provide a working space for their student ombudsperson, co-finance their activities, and provide administrative and technical support. The national students' council, called Croatian Students' Council (CSC), appoints a national Coordinator of Student Ombudspersons who coordinates the work of all HEI student ombudspersons, as described in the CSC's Rules of Procedure (Poslovnik o Radu Hrvatskog Studentskog Zbora, 2014). Together, the HEI/faculty policies on academic integrity (CoE, DC) and the Statute of the student council at the HEI/faculty level make up the academic integrity policies on an institutional level.

Analysis of Croatian Public Universities' Policies and Practices: Academic Integrity and Artificial Intelligence

Description of Institutional Policies at Analysed Public Universities in Croatia

Disciplinary bodies for employees use the Disciplinary Code for Employees (DCE) as their relevant document. DCEs describe employees' work-related obligations and types of misconduct; misconducts are usually ranked according to severity (minor, major and severe misconducts), along with descriptions of the types of misconduct. Depending on the level of severity, penalties are determined by the Disciplinary Committees. The only universities not explicitly ranking the severity of misconduct were in Pula (Juraj Dobrila University of Pula, 2023) and Slavonski Brod (University of Slavonski Brod, 2023). Examples of minor misconduct include not performing the duties according to the working contract, falsifying information, and disrupting processes of teaching or other institutional processes. Major misconduct examples include not respecting the general acts and decisions made by the University and its sub-units, misuse of position, gross negligence for work-related responsibilities, inappropriate or offensive behaviour toward students, discrimination, and corruption. Sanctions are determined according to the misconduct severity. Sanctions for minor misconducts include warnings, whilst major misconducts may be regular or extraordinary dismissal of the employee (at all Universities) or a financial compensation to the HEI if damage was caused (University North, 2014b). The Universities in Dubrovnik, Zagreb and Split allow individual Faculties to draw-up their own DCEs (Statut Sveučilišta u Splitu, 2023; Statut Sveučilišta u Zagrebu, 2023; Statute of the University of Dubrovnik, 2022). According to the Statute of the University of Split, the Senate should adopt the DCE for University employees, but this document was not available at the webpages of the university at the time of this research. Because I evaluated only institutional-based practices, the differences in effectiveness of university-based vs Faculty-based DCE policies could not be assessed within the scope of this research. Nonetheless, the existence of regulations, even when Faculties have higher autonomy in managing disciplinary procedures, points to the fact that the Universities of Zagreb and Split ensured that academic integrity policies are in place across the institution.

For students, the Disciplinary Code for Students (DCS) describes the types of misconduct (minor or major) and related consequences (Juraj Dobrila University of Pula, 2015; University North, 2014c; University of Dubrovnik, 2006b; University of Osijek, 2010; University of Rijeka, 2014; University of Split, 2010; University of Zadar, 2015). Major offenses include criminal acts, destruction of the university's property, falsification of grades and signatures of teaching staff, falsification of documents or information, and cheating or collaborating during exams. Minor offenses include the disruption of teaching processes or exams, failing to settle any obligations toward library services, and using university spaces for advertising, promotion, sales or gatherings unauthorised by the university, among others. In Zagreb, either the university or the Faculties could adopt DCSs (Statut Sveučilišta u Zagrebu, 2023), while in Split they were solely the responsibility of Faculties (Statut Sveučilišta u Splitu, 2023). At the University of Zagreb, the only DCS in place refers to the misuse of the student identification card, but no separate document was drawn for other types of misconduct (University of Zagreb, 2015).

Codes of Ethics (CoEs) at the public universities are used by the Ethics Committees and include: principles of academic freedom; the responsibilities of the teaching staff, students and employees; and descriptions of professional behaviour and professional misconduct (Josip Juraj Strossmayer University of Osijek, 2011; Juraj Dobrila University of Pula, 2008; Rukavina & Prijić-Samaržija, 2006; University North, 2014a; University of Dubrovnik, 2006a; University of Slavonski Brod, 2020; University of Split, 2009; University of Zadar, 2010; University of Zagreb, 2007). Misconduct includes plagiarism (except at the University North), accepting gifts and/or services in exchange for favours (except in Split and Zadar), authorship infringement, fabrication and falsification, and conflict of interest. Most universities, except those in Split and Zadar, also required that teaching staff be aware of new scientific achievements and developments in their fields of research, to be able to continuously improve.

Analysis of Institutional Practices: Academic Integrity

As mentioned in the Methodology section, I only analysed the documents made available on the universities' webpages. The Statutes[2] of all public universities prescribe the fundamental values of higher education, such as human rights, integrity and dignity of each person (Statut Sveučilišta Josipa Jurja Strossmayera u Osijeku (Nacrt Prijedloga), 2023; Statute of the University of Dubrovnik, 2022), ethical conduct (Statut Sveučilišta Sjever, 2023; Statut Sveučilišta u Zadru (Pročišćeni

[2] In Europe, a Higher Education Institution's (HEI's) Statute is a document that sets a constitutional framework of an institution, providing information on its governance, powers, responsibilities, members and/or other bodies and subjects aligned with the mission of the HEI.

Tekst), 2005), and the formation of disciplinary bodies. Only in the Statute of the University of Rijeka was the term academic integrity used (Statut Sveučilišta u Rijeci, 2023). At most universities, two sets of disciplinary bodies exist: one for students and one for employees and faculty members; in others, a single disciplinary body looks after students, staff and faculty members. Aside from the disciplinary bodies, each University has their own Ethics committee, dealing with ethical misconduct. These are the bodies dealing with academic integrity topics, so their practices are described below.

No university allowed students to be the members of employee disciplinary committees, except at the University North if a student reported academic misconduct by an employee (University North, 2014b). Students have the right to report employee misconduct at all universities, except at the University of Zadar (2006). Even though all universities had operating disciplinary committees, in some cases, if misconduct was classified as 'minor offense', the Dean, Vice-Rector or Rector could unilaterally determine the disciplinary measure (University of Osijek, 2018; University of Zadar, 2006). At some universities, like Rijeka and Slavonski Brod, a procedure for complaints in disciplinary procedures for employees was not in place (University of Rijeka, 2023b; University of Slavonski Brod, 2023), while at the University of Osijek, a disciplinary hearing was absent for decisions made about minor misconduct (University of Osijek, 2018). For students, decisions were made by Disciplinary Committees for Students at those universities with a DCS (Dubrovnik, North, Osijek, Pula, Rijeka Zadar, Zagreb). Unlike other universities, at the University of Zadar two out of three members of the disciplinary committee for students were students, giving the student population more autonomy over the matters of academic misconduct. Potential sanctions for students' misconduct, common to analysed universities, are described in Table 10.1 below.

Academic Integrity was specifically included in the CoEs of the universities of Zadar Rijeka, North, and Pula (Juraj Dobrila University of Pula, 2008; University North, 2014a; University of Rijeka, 2019; University of Zadar, 2010). Most universities had permanent Ethical Committees at the institutional level (see Table 10.2). At the University of Rijeka, there was a CoE for Students which was a unique document to help students in understanding students' ethical conduct (University of Rijeka, 2019). In this document, academic integrity and academic misconduct were explicitly mentioned. More information on the individual academic integrity bodies at each analysed public university is available in Table 10.2. However, there were no specific mentions of academic misconduct related to the use of AI or GenAI in any of the analysed procedures or policies at the universities (Table 10.3).

The information available on the webpages of Students' Councils was scarce, with only the University of Rijeka Student Council reporting on the Plan of Work and the activities of the University's student ombudsperson. In the University of Rijeka Students' Council's Plan of Work, one of the focus points was the increase in digital competencies of students; other initiatives, such as the formation of a body responsible for considering whether disciplinary procedures against students, are organized on a case-to-case basis (University of Rijeka Students' Council, 2022). To date, there were no publicly announced projects by Students Councils on the use of GenAI in higher education or the opportunities and threats related to it. However, because students are involved in the governance structures at universities, their

Table 10.1 Common sanctions for minor and major offenses at the analysed universities in Croatia

Common academic integrity sanctions according to Disciplinary Codes for Students (DCSs)			
Disciplinary measure	Level of offense the measure is given for	Notes	Universities applying the measure
Warning: verbal or written	Minor	Osijek: gravity of offense not established	Dubrovnik, Osijek, Pula, Rijeka, Zadar, Zagreb
Final warning before expulsion	Minor	North: applies also to minor offenses; Zadar: applies only to major offenses	Dubrovnik, North, Zadar
Suspension from classes and/or exams	Major	North: applies also to minor offenses; Osijek: gravity of offense not established; Pula, Zadar: applies only to minor offenses	Dubrovnik, North, Osijek, Pula, Rijeka, Zadar
Temporary expulsion from studies	Major	North: applies also to minor offenses; Osijek: gravity of offense not established	Dubrovnik, North, Osijek, Pula, Rijeka
Permanent expulsion from studies	Major	Osijek: gravity of offense not established	Dubrovnik, Osijek, Pula, Rijeka, Zadar
Temporary taking away student's identification card	Gravity of offense not established	/	Zagreb

Table 10.2 General overview of academic integrity bodies at the public universities in Croatia, at the institutional level (faculty-level information is not included). Only bodies for which documents were available on universities' web pages are described. *CoE* Code of Ethics, *DCE* Disciplinary Code for Employees, *DCS* Disciplinary Code for Students

University bodies: Academic Integrity					
University	Type of body	Body formation	Relevant document	Composition of the body	Notes
Dubrovnik	Ethics Committee	Permanent	CoE, Rules of Procedure of the Ethics Committee	5 members (4 teachers and 1 student)	Mandate: 4 years Who initiates the process after reporting: Ethics Committee
	Disciplinary Court	Ad hoc (at the department student attends)	DCS	The Disciplinary Court is composed of 2 councils: (i) First-degree council: 3 members (2 teachers, 1 student); (ii) Second-degree council: Department council	Who initiates the process after reporting: Department Head The Court only deals with the disciplinary procedures involving students

(continued)

Table 10.2 (continued)

University bodies: Academic Integrity					
University	Type of body	Body formation	Relevant document	Composition of the body	Notes
North	Ethics Committee	Permanent	CoE of the University North, CoE for scientific research of the University North	Composition is determined by the University Senate, based on the proposal by the Rector, and it has to include both the teachers and the students	Mandate: 3 years Who initiates the process after reporting: Rector Ethical Committee gives an opinion on the ethical nature of requests and submits their report to the Rector (or Vice-Rector) and to the person who filed a report
	Disciplinary Committee for Students	Ad hoc	DCS	3 members: 2 teachers, 1 student	Who initiates the process after reporting: Rector For minor offenses, shortened procedure without disciplinary hearing applies, and the Rector decides on a disciplinary measure. For major offenses, the disciplinary hearing is held, and the Committee decides on a disciplinary measure
	Disciplinary Committee for Employees	Permanent	DCE	3 members: 2 teachers, 1 student	Mandate: 3 years Who initiates the process after reporting: President of the Disciplinary Committee for Employees The Disciplinary Committee for Employees acts as the first-degree council. The second-degree council is composed of the Rector's Collegium (members are Rector, Vice-Rectors, Rector's assistants etc.)

(continued)

Table 10.2 (continued)

University bodies: Academic Integrity					
University	Type of body	Body formation	Relevant document	Composition of the body	Notes
Osijek	Ethics Committee	Permanent	CoE	7 members: 4 are proposed by the Senate, 1 by the Rector, 1 by the University Council and 1 by the Students' Council	Mandate: 4 years Who initiates the process after reporting: Ethics Committee
	Disciplinary Committee for Students	Permanent	DCS	5 members: 3 teachers, 2 students	Mandate: 3 years for teachers. Students' mandate is determined according to the Law Who initiates the process after reporting: Dean/Department Head
	Disciplinary Committee for Employees	Ad hoc	DCE	3 members: all teachers	Who initiates the process after reporting: Dean/Department Head For both minor and major offenses, the Dean/Department Head decides on a disciplinary measure; however, for minor offenses, no disciplinary hearing is held. For major offenses, the decision of the Dean is based on the opinion of the Disciplinary Committee for Employees, after the disciplinary hearing

(continued)

Table 10.2 (continued)

University bodies: Academic Integrity					
University	Type of body	Body formation	Relevant document	Composition of the body	Notes
Pula	Ethics Committee	Permanent	CoE, Rules of Procedure of the Ethics Committee	Committee is composed of 1member from each Faculty belonging to the University, 1 student member and 1 staff member	Mandate: 4 years Who initiates the process after reporting: Ethics Committee Ad hoc Ethics Committee can be formed based on the opinion of the University's Ethics Committee Report can be filed even by the persons not directly involved in ethical misconduct. Ethical Committee gives an opinion on the ethical nature of requests and submits their report to the Rector and to the person who filed a report
	Disciplinary Committee for Students	Permanent	DCS	5 members: 3 teachers, 1 student, 1 person of legal background	Mandate: 2 years Who initiates the process after reporting: Disciplinary Committee for Students The Disciplinary bodies for students' misconduct in Pula are the Disciplinary Committee for Students and the Complaints Disciplinary Committee (composed of 3 members: Vice-rector for studies and students, Head of Secretariat at the University, President of the Student Council)
	Disciplinary Committee for Employees	Permanent	DCE	7 members: 4 teachers, 3 staff members	Mandate: 5 years. Who initiates the process after reporting: Rector 3 out of 7 members participate in a single Disciplinary procedure

(continued)

Table 10.2 (continued)

University bodies: Academic Integrity					
University	Type of body	Body formation	Relevant document	Composition of the body	Notes
Rijeka	Ethics Committee	Permanent	CoE	5 members: 4 employees (teachers, staff members),1 student	Mandate: 3 years Who initiates the process after reporting: Ethics Committee
	Council of Honour	Permanent	CoE	5 members: 4 teachers with the title of professor, 1 student	Mandate: 3 years Council of Honour interprets the CoE, gives opinions on ethical matters and gives opinion on the ethical misconduct of the university/Faculty heads
	Disciplinary Committee for Students	Permanent	DCS	3 members: 2 teachers, 1 student	Mandate: 3 years Who initiates the process after reporting: Disciplinary Committee for Students The Disciplinary bodies for students' misconduct in Rijeka are the Disciplinary Committee for Students, the Complaints Disciplinary Committee (5 members: 4 teachers, 1 student) and the Secretaries of these two bodies
	Disciplinary Committee for Employees	Permanent	DCE	5 members: 4 teachers, 1 staff member	Mandate: 3 years Who initiates the process after reporting: Employer (university or Faculty)
Slavonski Brod	Ethics Committee	Permanent	CoE	5 members: 3 proposed by the Senate, 1 by the Rector and 1 by the University Students' Council	Mandate: 4 years Who initiates the process after reporting: Ethics Committee
	Disciplinary Committee for Employees	Permanent	DCE	3 members: 2 teachers, 1 staff member	Mandate: 3 years Who initiates the process after reporting: Rector The Disciplinary Committee for Employees makes a decision based on the documentation received and the Disciplinary hearing

(continued)

Table 10.2 (continued)

University bodies: Academic Integrity					
University	Type of body	Body formation	Relevant document	Composition of the body	Notes
Split	Ethics Committee	Permanent	CoE	7 members: 4 staff members proposed by the Senate, 2 external members proposed by the University Council, 1 student proposed by the University Students' Council	Mandate: 4 years Who initiates the process after reporting: Ethics Committee
Zadar	Ethics Committee	Permanent	CoE, Rules of Procedure of the Ethics Committee	5 members: 3 teachers, 1 student, 1 staff member	Mandate: 4 years Who initiates the process after reporting: Ethics Committee
	Disciplinary Committee for Students	Permanent	DCS	3 members: 1 teacher, 2 students	Mandate: 2 years Who initiates the process after reporting: Disciplinary Committee for Students The Disciplinary bodies for students' misconduct in Zadar are the Disciplinary Committee for Students and the Complaints Disciplinary Committee (composed of 3 members: Vice-rector for studies and students; Secretary in charge for administration, human resources and studies; President of the Student Council)
	Disciplinary Committee for Employees	Permanent	DCE	3 members: 2 teachers, 1 staff member	Mandate: 3 years Who initiates the process after reporting: Rector

(continued)

Table 10.2 (continued)

University bodies: Academic Integrity					
University	Type of body	Body formation	Relevant document	Composition of the body	Notes
Zagreb	Ethics Committee	Ad hoc/ Permanent (formed as needed)	CoE	Not defined.	Who initiates the process after reporting: Senate or Ethics Council Ethics Committee is appointed by the Rector, based on the nominations from the Ethics Council If the body is formed on permanent basis (due to the non-existence of appropriate ethics committees at the faculties), the mandate of members is maximum 4 years
	Ethics Council	Permanent	CoE	9 members: most teachers, at least 1 student, at least 1 staff member	Mandate: 4 years Purpose: is provides counsel on general questions regarding ethical behaviour, it collects and updates information on the matters of ethics and it acts according to the CoE. The Ethics Council also deals with complaints on the procedures conducted by the Ethics Committee
	Disciplinary Committee for Students	Permanent	DCS	3 members: 2 teachers, 1 student	Mandate: 4 years Who initiates the process after reporting: Disciplinary Committee for Students This DCS applies only to the misuse of the student identity card

Table 10.3 List of institutional policies analysed per university, and the policies mentioning academic integrity and generative-AI

University	Location	Relevant policies	Policies mentioning academic integrity	Policies mentioning generative-AI
University of Dubrovnik	Dubrovnik	Statute, Code of Ethics, Disciplinary Code for Students, Statute of the Students' Council	None	None
University North	Varaždin	Statute, Code of Ethics, Disciplinary Code for Employees, Disciplinary Code for Students, Statute of the Students' Council	Code of Ethics	None
Josip Juraj Strossmayer University of Osijek	Osijek	Statute, Code of Ethics, Disciplinary Code for Employees, Disciplinary Code for Students, Statute of the Students' Council	None	None
Juraj Dobrila University of Pula	Pula	Statute, Code of Ethics, Disciplinary Code for Employees, Disciplinary Code for Students, Statute of the Students' Council	Code of Ethics	None
University of Rijeka	Rijeka	Statute, Code of Ethics, Code of Ethics for Students, Disciplinary Code for Employees, Disciplinary Code for Students, Statute of the Students' Council, Policy for the Use of AI Tools at the University of Rijeka	Statute, Code of Ethics (research integrity is mentioned), Code of Ethics for Students	Policy for the Use of AI Tools at the University of Rijeka
University of Slavonski Brod	Slavonski Brod	Statute, Code of Ethics, Disciplinary Code for Employees, Statute of the Students' Council	None	None
University of Split	Split	Statute, Code of Ethics, Statute of the Students' Council	None	None
University of Zadar	Zadar	Statute, Code of Ethics, Disciplinary Code for Employees, Disciplinary Code for Students, Statute of the Students' Council	Code of Ethics	None
University of Zagreb	Zagreb	Statute, Code of Ethics, Disciplinary Code for Students, Statute of the Students' Council	None	None

opinions and activities may prove valuable for the universities beyond the student councils. These activities could further contribute to the development of students' awareness about the proper use of GenAI.

Analysis of Institutional Strategies: Academic Integrity and Generative AI (GenAI)

As no specific mention of GenAI was found in academic integrity documents at any of the universities (Table 10.3), I instead evaluate universities strategic documents to determine if any mention of it could indicate universities' awareness of the rise in the use of GenAI. For instance, the University of Dubrovnik's 2020–2025 research strategy did not mention academic integrity or any changes to current ethical procedures to accommodate the use of GenAI (University of Dubrovnik, 2020). In the University North's Scientific Research and Artistic Strategy for the period 2021–2027, the use of GenAI is mentioned in fields such as medical, technical and biotechnical sciences, but no explicit links were made between the academic integrity and GenAI (University North, 2021). At the Universities of Osijek and Rijeka, GenAI (or AI generally) were not mentioned (Josip Juraj Strossmayer University of Osijek, 2021; University of Rijeka, 2021). At the University of Pula, strengthening research integrity was included as part of scientific and research excellence. However, these activities mainly focused on the development of research infrastructure and the recognition of scientists (Juraj Dobrila University of Pula, 2021). In the Strategies of the universities of Slavonski Brod and Split, the use of AI was mentioned, but only in relation to Engineering courses (Slavonski Brod) or as a building block of the university's digitalisation plan (Split, Zagreb) (University of Slavonski Brod, 2021; University of Split, 2021; University of Zagreb, 2014). The University of Split issued a Research Strategy 2022–2026 in which the culture of openness and research integrity was described as being further strengthened and aligned with the European Code of Conduct for Research Integrity, despite the lack of key performance indicators regarding changes in policies related to research (University of Split, 2022). The University of Zadar did not to have a publicly available Strategy for the period after 2022.

An example of a positive practice in addressing the use of GenAI came from the University of Rijeka, where the Council of Honour issued a Communication on Responsible use of Artificial Intelligence Tools in April 2023 (University of Rijeka, 2023c) based on the position paper of the European University Association titled "Artificial intelligence tools and their responsible use in higher education learning and teaching" (European University Association, 2023). In the Communication, students were told they should always reference any use of GenAI software and that all ethical principles applied in teaching and learning, even when the use of GenAI is required. Additionally, in 2024, the University published a Policy for the Use of AI Tools where the advantages and risks for the use of AI tools were described

(University of Rijeka, 2024). In this policy, general definitions of terms, including "generative AI" were given; it was highlighted that the use of AI tools must be based on research and teaching excellence, in accordance with the principles of academic integrity. Some of the risks related to the use of AI tools included lack of transparency of information and results; increased possibility of plagiarism, data manipulation and fabrication; lack of data reproducibility; data and algorithm biasness; possibility of spreading false information; lack of clarity regarding intellectual property rights; data privacy and safety issues; and contributing to social inequalities.

Despite the analysed policies not reflecting on the use of GenAI broadly, it was possible to find multiple media sources pointing to the use of AI in higher education (in initiatives and courses) at public universities. For example, the Days of e-learning at the University of Rijeka (University of Rijeka, 2023a); the project 'transversal skills in applied artificial intelligence (TSAAI) (Načinović, 2022), the project 'Artificial Intelligence in Public Relations (AI in PR) (Srednja.hr, 2021); and the free online course 'Elements of AI' in collaboration with the University of Zagreb (MinnaLearn & University of Helsinki, n.d.; University of Zagreb, 2021), among others. The Croatian Artificial Intelligence Association (CroAI) was established as well, which is one of the founding members of The European AI Forum (Croatian Artificial Intelligence Association, 2021). Additionally, student-led media reported on the threats of using GenAI, finding that several interviewed faculties apply anti-plagiarism tools with little success in detecting AI-generated text (Markovinović, 2023), suggesting that further education of teaching staff and students is necessary in terms of misconduct related to the use of GenAI tools.

Discussion and Conclusions

Croatia's public universities tend to adopt policies on academic integrity based on national legislation and in alignment with the European policies, leaving a broad space for improvement especially regarding the opportunities and threats of digital learning environments. University policies differed based on the composition and the jurisdiction of bodies (either at the university or at the faculty level), as well as on defining misconduct and students' involvement in the disciplinary processes (Table 10.2). Findings from the SEEPPAI project continue to apply to the Croatian Higher Education System (Lancaster et al., 2017). All public universities have emphasised the need for improvements in the field of digitalization, including digital transitions and opening new e-learning modules or even programmes (Josip Juraj Strossmayer University of Osijek, 2021; Juraj Dobrila University of Pula, 2021; University North, 2021; University of Dubrovnik, 2020; University of Rijeka, 2021; University of Slavonski Brod, 2021; University of Split, 2021; University of Zadar, 2020; University of Zagreb, 2014). The universities of Slavonski Brod and Split include the application of GenAI in teaching and learning, while in Zadar and Zagreb the digitalization of learning materials was forefront (University of Zadar, 2020; University of Zagreb, 2014).

The European Parliament's 2021 Resolution on Artificial Intelligence in Education, Culture and the Audiovisual Sector (European Parliament, 2021) underlines the importance of GenAI and the need for its ethical and respectful use regarding fundamental rights, freedoms and values. Explicitly mentioned was the idea that the teachers and students who use AI in education must be well-trained. Additionally, high levels of digital literacy are a prerequisite for the use of GenAI in education. The European Parliament called upon the European Commission to propose a legal framework for the use of GenAI. A 'Proposal for a Regulation of the European Parliament and of the Council laying down harmonised rules on artificial intelligence' was created (European Commission, 2021), passed and published by July 2024 (European Parliament & The Council of the European Union, 2024); it may require other legislative changes within the European Union, and will certainly bring changes to Croatian Higher Education policies.

Croatian public universities generally have not paid adequate attention to the rising use of GenAI, possibly because they did not perceive the use of it as a potential threat to HE. However, on a general level, the importance of advancing students' digital skills, offering online courses and programmes, digitalizing processes and even applying AI in teaching and learning was evident. Despite their planned developments in digitalization, none of the universities linked those digitalization strategies to the importance of having effective policies on academic integrity (either through CoEs or DCS). Even more concerning is that the University of Dubrovnik had not updated their academic integrity policies (or uploaded the new versions of documents for public viewing) in more than 15 years; this could prove problematic in identifying new threats, especially those arising from the use of GenAI. Student ombudspersons had limited public roles in the prevention of academic misconduct. At the University of Rijeka, the Coordinator of student ombudspersons (appointed by the Students' Council) did not conduct any programmes or projects related to the use of GenAI in higher education. Generally, students seemed uninvolved in independent initiatives regarding academic integrity or the use of GenAI, despite appointing students in the bodies dealing with ethical matters or disciplinary bodies. Only one example of good practice was identified at the University of Rijeka, where the Council of Honour communicates to students that GenAI in teaching and learning should be referenced and used in ethical and transparent ways, and that its misuse could lead to disciplinary sanctions. The University of Rijeka also created the first Croatian HEI higher education policy prescribing the safe use of AI tools, including GenAI, in teaching, learning and research for all members of the academic community. However, current policies related specifically to academic integrity (CoEs, DCE, DCS) at the University of Rijeka do not reference academic misconduct related to the use of AI.

Overall, the policies at Croatian public universities need to be updated to reflect current digitalisation measures and plans regarding the use of digital technologies and GenAI. Furthermore, public universities should strive to educate their teachers and learners on the ethical use of GenAI and should build academic integrity values rather than simply prescribing punitive measures for academic misconduct.

References

Bajić, B., Đidara, V., Vlahoviček, K., Buljan, H., Gruber, L., Jakšić, M., & Podnar, A. (2023). *Analitičko izvješće o provedenom znanstvenom i tehnologijskom mapiranju* (pp. 1–97). Ministry of Science and Education of the Republic of Croatia. https://mzo.gov.hr/UserDocsImages/dokumenti/EUfondovi/OPKK_2014-2020/ZTP/Analiticko-izvjesce-mapiranje-umjetna-inteligencija.pdf

Bašić, Ž., Kružić, I., Jerković, I., Buljan, I., & Marušić, A. (2019). Attitudes and knowledge about plagiarism among university students: Cross-sectional survey at the University of Split, Croatia. *Science and Engineering Ethics, 25*(5), 1467–1483. https://doi.org/10.1007/s11948-018-0073-x

Bašić, Ž., Banovac, A., Kružić, I., & Jerković, I. (2023a). Better by you, better than me, chatgpt3 as writing assistance in students essays. *arXiv Preprint arXiv:2302.04536*. https://doi.org/10.48550/arXiv.2302.04536

Bašić, Ž., Banovac, A., Kružić, I., & Jerković, I. (2023b). ChatGPT-3.5 as writing assistance in students' essays. *Humanities and Social Sciences Communications, 10*(1), 750. https://doi.org/10.1057/s41599-023-02269-7

Borovečki, A., Ten Have, H., & Orešković, S. (2004). Developments Regarding Ethical Issues in Medicine in the Republic of Croatia. *Cambridge Quarterly of Healthcare Ethics, 13*(3), 263–266. https://doi.org/10.1017/S0963180104133094

Čepo, D. (2017). *Političko-pravni mehanizmi sprečavanja neetičnih praksi u hrvatskoj znanosti i visokom obrazovanju: Mapiranje problema i mogućih rješenja* (pp. 1–18). https://oz.goo.hr/wp-content/uploads/2017/10/2017_OZ_Neeticne-prakse-u-znanosti-i-obrazovanju.pdf

Committee for Ethics in Science and Higher Education. (2006). *Etički kodeks odbora za etiku u znanosti i visokom obrazovanju*. Ministry of Science and Education of the Republic of Croatia. https://mzo.gov.hr/UserDocsImages/dokumenti/StrucnaTijela/Eti%C4%8Dki%20kodeks%20odbora%20za%20etiku%20u%20znanosti%20i%20visokom%20obrazovanju.pdf

Croatian Artificial Intelligence Association. (2021). Croatian Artificial Intelligence Association. https://www.croai.org/

Croatian Bureau of Statistics. (2023). *Studenti Upisani na Stručni i Sveučilišni Studij u Zimskom Semestru Ak. G. 2022./2023.* (OBR-2023-4-5) [Dataset]. https://podaci.dzs.hr/2023/hr/58225

Eke, D. O. (2023). ChatGPT and the rise of generative AI: Threat to academic integrity? *Journal of Responsible Technology, 13*, 100060. https://doi.org/10.1016/j.jrt.2023.100060

ENQA, ESU, EUA, & EURASHE. (2015). *Standards and Guidelines for Quality Assurance in the European Higher Education Area (ESG)*. https://www.enqa.eu/wp-content/uploads/2015/11/ESG_2015.pdf

European Commission. (2021). Proposal for a Regulation of the European Parliament and of the Council laying down harmonised rules on artificial intelligence. European Commission. https://eur-lex.europa.eu/legal-content/EN/TXT/PDF/?uri=CELEX:52021PC0206

European Parliament. (2021). *Izvješće o umjetnoj inteligenciji u obrazovnom, kulturnom i audiovizualnom sektoru*. European Parliament. https://www.europarl.europa.eu/doceo/document/A-9-2021-0127_HR.html

European Parliament, & The Council of the European Union. (2024, July 12). *Regulation (EU) 2024/1689 of the European Parliament and of the Council of 13 June 2024 laying down harmonised rules on artificial intelligence and amending Regulations (EC) No 300/2008, (EU) No 167/2013, (EU) No 168/2013, (EU) 2018/858, (EU) 2018/1139 and (EU) 2019/2144 and Directives 2014/90/EU, (EU) 2016/797 and (EU) 2020/1828 (Artificial Intelligence Act) Text with EEA relevance*. Official Journal of the European Union. https://eur-lex.europa.eu/legal-content/EN/TXT/?uri=OJ:L_202401689

European University Association. (2023). *Artificial intelligence tools and their responsible use in higher education learning and teaching*. European University Association. https://eua.eu/resources/publications/1059:artificial-intelligence-tools-and-their-responsible-use-in-higher-education-learning-and-teaching.html

Eurostat. (2020). Do young people in the EU have digital skills? [News Article]. https://ec.europa.eu/eurostat/web/products-eurostat-news/-/edn-20200715-1#:~:text=Among%20EU%20Member%20States%2C%20Croatia,well%20as%20Greece%20(92%25)

Foltynek, T., Bjelobaba, S., Glendinning, I., Khan, Z. R., Santos, R., Pavletić, P., & Kravjar, J. (2023). ENAI Recommendations on the ethical use of Artificial Intelligence in Education. *International Journal for Educational Integrity, 19*(1), 12. https://doi.org/10.1007/s40979-023-00133-4

Glendinning, I., Foltýnek, T., Dlabolová, D., Linkeschová, D., & Lancaster, T. (2017). Exploring issues challenging academic integrity in South East Europe. In *Plagiarism across Europe and beyond*, pp. 132–146.

Josip Juraj Strossmayer University of Osijek. (2011). *Etički kodeks Sveučilišta Josipa Jurja Strossmayera u Osijeku.* Josip Juraj Strossmayer University of Osijek. http://www.unios.hr/wp-content/uploads/2015/07/SJJS_Eticki_kodeks.pdf

Josip Juraj Strossmayer University of Osijek. (2021). *Strategija Sveučilišta Josipa Jurja Strossmayera u Osijeku 2021.–2030.* Josip Juraj Strossmayer University of Osijek. http://www.foozos.unios.hr/dokumenti/kvaliteta/dokumenti/na%20razini%20sveu%C4%8Dili%C5%A1ta/Strategija-Sveu%C4%8Dili%C5%A1ta-2021-2030.pdf

Juraj Dobrila University of Pula. (2008). *Etički kodeks Sveučilišta Jurja Dobrile u Puli.* Juraj Dobrila University of Pula. https://www.unipu.hr/_download/repository/Eticki_kodeks_Sveucilista_Jurja_Dobrile_u_Puli.pdf

Juraj Dobrila University of Pula. (2015). *Pravilnik o stegovnoj odgovornosti studenata.* Juraj Dobrila University of Pula. https://www.unipu.hr/_download/repository/Studentska_pitanja/Pravilnik_o_stegovnoj_odgovornosti_studenata_2017.pdf

Juraj Dobrila University of Pula. (2021). *Strategija razvoja Sveučilišta Jurja Dobrile u Puli 2021.–2026.* Juraj Dobrila University of Pula. https://www.unipu.hr/_news/93569/Strategija%20razvoja%20Sveucilista%202021.%20-%202026..pdf

Juraj Dobrila University of Pula. (2023). *Pravilnik o stegovnoj odgovornosti zaposlenika Sveučilišta Jurja Dobrile u Puli.* Juraj Dobrila University of Pula. https://www.unipu.hr/_download/repository/Pravilnik%20o%20stegovnoj%20odgovornosti%20zaposlenika%20UNIPU.pdf

Katavić, I., Budimir Šoško, G., & Kopecki, D. (2016). Razvoj kulture akademskog integriteta u online obrazovnom okruženju. *Obrazovanje Za Poduzetništvo – E4E : Znanstveno Stručni Časopis o Obrazovanju Za Poduzetništvo, 6*(1), 111–120.

Kern, J., & Petrovečki, M. (2009). Medicinsko odlučivanje. In *Medicinska informatika* (pp. 179–197). Medicinska Naklada.

Lancaster, T., Foltynek, T., Glendinning, I., & Henek Dlabolova, D. (2017). *South East European Project on Policies for Academic Integrity final report* (pp. 1–58) [Final Report]. Council of Europe Pan-European Platform on Ethics, Transparency and Integrity in Education (ETINED).

Malički, M., Katavić, V., Marković, D., Marušić, M., & Marušić, A. (2019). Perceptions of ethical climate and research pressures in different faculties of a university: Cross-sectional study at the University of Split, Croatia. *Science and Engineering Ethics, 25*(1), 231–245. https://doi.org/10.1007/s11948-017-9987-y

Markovinović, M. (2023). *Fakulteti u Hrvatskoj znaju za chatbot koji piše i radove: Pitali smo ih kako se planiraju 'boriti' protiv njega.* Srednja.Hr. https://www.srednja.hr/svastara/fakulteti-u-hrvatskoj-znaju-za-chatbot-koji-pise-i-radove-pitali-smo-ih-kako-se-planiraju-boriti-protiv-njega/

Milinković, A., Vuleta, D., & Babić, I. T. (2024). Student perception of using generative AI tools in relation to academic integrity and their advantages and disadvantages. In *2024 47th MIPRO ICT and Electronics Convention (MIPRO)*, pp. 601–606. https://doi.org/10.1109/MIPRO60963.2024.10569274

Ministry of Science and Education of the Republic of Croatia. (2016). *Okvir nacionalnoga kurikuluma.* Ministry of Science and Education of the Republic of Croatia. http://www.kurikulum.hr/wp-content/uploads/2016/02/ONK-18.2-POPODNE-2.pdf

MinnaLearn, & University of Helsinki. (n.d.). *Elements of AI* [Online Course]. https://www.elementsofai.com/hr

Moodle. (2019). *Virtual Learning Environment built on Moodle becomes an integral part of teaching and learning processes in higher education in Croatia*. (pp. 1–4). University of Zagreb. https://moodle.com/wp-content/uploads/2019/07/University-of-Zagreb_Casestudy.pdf
Načinović, A. (2022). Tehnički fakultet u Rijeci partner na projektu koji razvija obrazovnu platformu o primjeni umjetne inteligencije u visokom obrazovanju. *Torpedo Media*. https://torpedo.media/novosti-rijeka/tehnicki-fakultet-u-rijeci-partner-na-projektu-koji-razvija-obrazovnu-platformu-o-primjeni-umjetne-inteligencije-u-visokom-obrazovanju
Poslovnik o Radu Hrvatskog Studentskog Zbora, Croatian Students' Council, 9 (2014). https://hsz.hr/wp-content/uploads//HSZ_Poslovnik_o_radu_Finalna_12_4_2014.pdf
Rukavina, D., & Prijić-Samaržija, S. (2006). *Etički kodeks Sveučilišta u Rijeci*. University of Rijeka. http://www.riteh.uniri.hr/media/filer_public/9d/e3/9de377fa-8902-4df2-8a3e-ea8de4411ac7/eticki_kodeks_sveucilista_u_rijeci_2006.pdf
Srednja.hr. (2021). Veliki projekt za studente: Naučite kako koristiti umjetnu inteligenciju u odnosima s javnošću. Srednja.Hr. https://www.srednja.hr/faks/veliki-projekt-za-studente-naucite-kako-koristiti-umjetnu-inteligenciju-u-odnosima-s-javnoscu/
Statut Sveučilišta Josipa Jurja Strossmayera u Osijeku (Nacrt Prijedloga), 1 (2023). http://www.unios.hr/wp-content/uploads/2023/01/NOVI-STATUT-01-2023-1.pdf
Statut Sveučilišta Sjever, 2137-0336-09-22–33 602-04/23-02/4 (2023).
Statut Sveučilišta u Rijeci, 1 (2023). https://uniri.hr/wp-content/uploads/2023/02/Nacrt-prijedloga-Statuta-Sveucilista-u-Rijeci_08_02_2023.pdf
Statut Sveučilišta u Splitu, 2181-202-3-01–3 029-03/23-01/03 1 (2023). https://www.unist.hr/DesktopModules/Bring2mind/DMX/API/Entries/Download?language=hr-HR&EntryId=2278&Command=Core_Download&PortalId=0&TabId=1846
Statut Sveučilišta u Zadru (Pročišćeni Tekst), AL-01-01-1206/1-2005 1 (2005). http://www.unizd.hr/Portals/0/doc/StatutUnizd.pdf
Statut Sveučilišta u Zagrebu, 251-25-03-04/1-23–5 011-02/23-02/02 1 (2023). http://www.unizg.hr/fileadmin/rektorat/O_Sveucilistu/Dokumenti_javnost/Dokumenti/Strateski_dokumenti/statut/Statut_SUZG_ZAVRSNO_fin_fin_travanj2023.pdf
Statute of the University of Dubrovnik, 1591-1/22 (2022). https://www.unidu.hr/wp-content/plugins/quarascope/download.php?file=30576
The European Commission. (2022). *Digital Economy and Society Index (DESI) 2022. Croatia* (pp. 1–18) [Country Reports]. https://digital-strategy.ec.europa.eu/en/policies/desi-croatia
The European Higher Education Area. (1999). *The Bologna Declaration of 19 June 1999*. The European Higher Education Area. https://www.ehea.info/media.ehea.info/file/Ministerial_conferences/02/8/1999_Bologna_Declaration_English_553028.pdf
University North. (2014a). *Etički kodeks Sveučilišta Sjever*. University North. https://www.unin.hr/wp-content/uploads/Eti%C4%8Dki-kodeks-Sveu%C4%8Dili%C5%A1ta-Sjever.pdf
University North. (2014b). *Pravilnik o stegovnoj i materijalnoj odgovornosti zaposlenika*. University North. https://www.unin.hr/wp-content/uploads/Pravilnik-o-stegovnoj-i-materijalnoj-odgovornosti-zaposlenika.pdf
University North. (2014c). *Pravilnik o stegovnoj odgovornosti studenta*. University North. https://www.unin.hr/wp-content/uploads/Pravilnik-stegovna-odgovornost-studenata.pdf
University North. (2021). *Scientific Research and Artistic Strategy of University North for the period 2021–2027*. University North. https://www.unin.hr/wp-content/uploads/Scientific-Research-and-Artistic-Strategy-of-University-North_2021_2027.pdf
University of Dubrovnik. (2006a). *Etički kodeks znanstvenika, nastavnika i suradnika Sveučilišta u Dubrovniku*. University of Dubrovnik. https://www.unidu.hr/wp-content/plugins/quarascope/download.php?file=2446
University of Dubrovnik. (2006b). Pravilnik o stegovnoj odgovornosti studenata Sveučilišta u Dubrovniku. https://www.unidu.hr/wp-content/plugins/quarascope/download.php?file=2448
University of Dubrovnik. (2020). *Strategija istraživanja Sveučilišta u Dubrovniku 2020–2025*. University of Dubrovnik. https://www.unidu.hr/wp-content/uploads/2021/04/Strategija-istrazivanja-Sveucilista-u-Dubrovniku-2020-2025_web.pdf

University of Osijek. (2010). *Pravilnik o stegovnoj odgovornosti studenata Sveučilišta Josipa Jurja Strossmayera u Osijeku*. University of Osijek.

University of Osijek. (2018). *Pravilnik o stegovnoj odgovornosti nastavnika i suradnika Sveučilišta Josipa Jurja Strossmayera u Osijeku* (003-05/18-01/4). University of Osijek. http://www.unios.hr/wp-content/uploads/2018/01/Pravilnik-2018.pdf

University of Rijeka. (2014). *Pravilnik o stegovnoj odgovornosti studenata Sveučilišta u Rijeci* (602-04/14-01/02; Version 2170-57-01-14-330). University of Rijeka. https://uniri.hr/wp-content/uploads/2019/03/Pravilnik_o_stegovnoj_odgovornosti_studenata_29_listopada_2014.pdf

University of Rijeka. (2019). *Etički kodeks za studentice/ studente Sveučilišta u Rijeci*. University of Rijeka. https://uniri.hr/wp-content/uploads/2019/05/Etic48dki20kodeks20za20studente20i20studentice.pdf

University of Rijeka. (2021). *University of Rijeka Strategy 2021–2025*. University of Rijeka.

University of Rijeka. (2023a). *Dani e-učenja na Sveučilištu u Rijeci*. https://uniri.hr/wp-content/uploads/2023/06/2023_Dani_e-ucenja_Program.pdf

University of Rijeka. (2023b). *Pravilnik o stegovnoj odgovornosti zaposlenika Sveučilišta u Rijeci* (030-01/23-01/9; Version 2170-137-01-23-1). University of Rijeka. https://uniri.hr/wp-content/uploads/2023/06/2023_05_23_Pravilnik_o_stegovnoj_odgovornosti_zaposlenika-1.pdf

University of Rijeka. (2023c). *Priopćenje Vijeća časti o odgovornoj uporabi alata umjetne inteligencije*. University of Rijeka. https://uniri.hr/o-sveucilistu/struktura/vijece-casti/

University of Rijeka. (2024, January). *Politika korištenja alata umjetne inteligencije na Sveučilištu uRijeci*. University of Rijeka. https://medri.uniri.hr/wp-content/uploads/2024/01/13.-Donosenje-Politike-koristenja-alata-umjetne-inteligencije-na-Sveucilistu-u-Rijeci.pdf

University of Rijeka Students' Council. (2022). *Plan i program rada Studentskog zbora Sveučilišta u Rijeci 2023*. University of Rijeka Students' Council. https://sz.uniri.hr/wp-content/uploads/2023/04/3.-Plan-i-program-rada-SZSUR-2023.pdf

University of Slavonski Brod. (2020). *Etički kodeks Sveučilišta u Slavonskom Brodu*. University of Slavonski Brod. https://unisbhr.sharepoint.com/sites/WebRepozitorij/Zajednicki%20dokumenti/Forms/AllItems.aspx?id=%2Fsites%2FWebRepozitorij%2FZajednicki%20dokumenti%2FDokumenti%2FAktiSveucilista%2FStatut%2FEticki%20kodeks%20Sveucilista%20u%20Slavonskom%20Brodu%2Epdf&parent=%2Fsites%2FWebRepozitorij%2FZajednicki%20dokumenti%2FDokumenti%2FAktiSveucilista%2FStatut&p=true&ga=1

University of Slavonski Brod. (2021). *Strategija razvoja Sveučilišta u Slavonskom Brodu za razdoblje 2021.–2027*. University of Slavonski Brod. https://unisbhr.sharepoint.com/sites/WebRepozitorij/Zajednicki%20dokumenti/Forms/AllItems.aspx?id=%2Fsites%2FWebRepozitorij%2FZajednicki%20dokumenti%2FDokumenti%2FAktiSveucilista%2FStrategija%2FStrategija%20razvoja%20UNISB%202021%2E%20final%2Epdf&parent=%2Fsites%2FWebRepozitorij%2FZajednicki%20dokumenti%2FDokumenti%2FAktiSveucilista%2FStrategija&p=true&ga=1

University of Slavonski Brod. (2023). *Pravilnik o stegovnoj i materijalnoj odgovornosti zaposlenika Sveučilišta u Slavonskom Brodu* (602-01/23-09/4; Version 2178-1-39-01-23-2). University of Slavonski Brod. https://unisbhr.sharepoint.com/sites/WebRepozitorij/Zajednicki%20dokumenti/Forms/AllItems.aspx?id=%2Fsites%2FWebRepozitorij%2FZajednicki%20dokumenti%2FDokumenti%2FAktiSveucilista%2FPravilnici%2FPravilnik%20o%20stegovnoj%20odgovornosti%20zaposlenika%20UNISB%2Epdf&parent=%2Fsites%2FWebRepozitorij%2FZajednicki%20dokumenti%2FDokumenti%2FAktiSveucilista%2FPravilnici&p=true&ga=1

University of Split. (2009). *Etički kodeks Sveučilišta u Splitu*. University of Split. https://www.unist.hr/DesktopModules/Bring2mind/DMX/API/Entries/Download?language=hr-HR&EntryId=124&Command=Core_Download&PortalId=0&TabId=1846

University of Split. (2010). *Pravilnik o stegovnoj odgovornosti studenata* (003-05/10-02/0001; Version 2181/206-01-10-0001). University of Split. https://data.fesb.unist.hr/public/documents/general_curriculum/7_Pravilnik%20o%20stegovnoj%20odgovornosti%20studenata_2010-03-18.pdf

University of Split. (2021). *Strategija Sveučilišta u Splitu 2021.–2025.* University of Split. http://ozs.unist.hr/hr/?preview=1&option=com_dropfiles&format=&task=frontfile.download&catid=160&id=1543&Itemid=1000000000000

University of Split. (2022). *Znanstvena strategija Sveučilišta u Splitu 2022.*–2026. University of Split. https://www.unist.hr/DesktopModules/Bring2mind/DMX/API/Entries/Download?language=hr-HR&EntryId=2177&Command=Core_Download&PortalId=0&TabId=1580

University of Zadar. (2006). *Pravilnik o stegovnoj odgovornosti radnika za povrede radne dužnosti (pročišćeni tekst)* (AL-01-01-531/1-2006). University of Zadar. http://www.unizd.hr/Portals/0/pdf/Pravilnik_o_stegovnoj_odgovornosti_radnika_za_povrede_radnih_duznosti.pdf

University of Zadar. (2010). *Etički kodeks*. University of Zadar. https://www.unizd.hr/Portals/0/doc/doc_pdf_dokumenti/ostali/eticki_kodeks_20100525.pdf

University of Zadar. (2015). *Pravilnik o stegovnoj odgovornosti studenata/ studentica (pročišćeni tekst)* (012-01/11-01/03; Version 2198-1-79-12/15-03). University of Zadar. https://www.unizd.hr/Portals/0/doc/doc_pdf_dokumenti/pravilnici/pravilnik_o_stegovnoj_odgovornosti_studenata_20150917.pdf

University of Zadar. (2020). *Strateški program znanstvenih istraživanja u području društvenih i humanističkih znanosti te u umjetničkom području od 2020.* Do 2024. University of Zadar. https://www.unizd.hr/Portals/0/doc/doc_pdf_dokumenti/strategije/strateski_program_unizd_drustveno_humanisticko_umjetnicko_2020_2024_20211011.pdf

University of Zagreb. (2007). *Etički Kodeks. Sveučilište u Zagrebu,* 2009. University of Zagreb. http://www.unizg.hr/fileadmin/rektorat/O_Sveucilistu/Dokumenti_javnost/Propisi/Pravilnici/Eticki_kodeks.pdf

University of Zagreb. (2014). *Strategija razvoja podrške studentima Sveučilišta u Zagrebu (2013.–2025.).* University of Zagreb. http://www.unizg.hr/fileadmin/rektorat/O_Sveucilistu/Dokumenti_javnost/Dokumenti/Strateski_dokumenti/Izvjesca/Strategija_razvoja_podrske__studentima.pdf

University of Zagreb. (2015). *Pravilnik o stegovnoj odgovornosti studenata za zlouporabe studentske iskaznice* (012-03/15-03/09; Version 380-020/173-15-1). University of Zagreb. http://www.unizg.hr/fileadmin/rektorat/O_Sveucilistu/Dokumenti_javnost/Dokumenti/Propisi/Pravilnik_o_stegovnoj_odgovornosti_studenata_za_zlouporabe_studentske_iskaznice.pdf

University of Zagreb. (2021). *Veliki uspjeh edukacijskoga programa 'Elementi umjetne inteligencije'.* University of Zagreb. https://www.unizg.hr/nc/vijest/article/veliki-uspjeh-edukacijskoga-programa-elementi-umjetne-inteligencije/

Zakon o osiguravanju kvalitete u visokom obrazovanju i znanosti, NN 151/2022, Croatian Parliament, 2330 6 (2022). https://narodne-novine.nn.hr/clanci/sluzbeni/2022_12_151_2330.html

Zakon o Studentskom Zboru i Drugim Studentskim Organizacijama, NN 71/2007, Croatian Parliament, 2182 (2007). /eli/sluzbeni/2007/71/2182.

Zakon o Visokom Obrazovanju i Znanstvenoj Djelatnosti, NN 119/2022, Croatian Parliament, 71-10-01/1-22-2 011-02/22-02/139 1834 (2022). https://narodne-novine.nn.hr/clanci/sluzbeni/2022_10_119_1834.html

The manufacturer's authorised representative in the EU is Springer Nature Customer Service Centre GmbH, Europaplatz 3, 69115 Heidelberg, Germany. If you have any concerns regarding our products, please contact ProductSafety@springernature.com

Printed and bound by CPI Group (UK) Ltd, Croydon, CR0 4YY
30/09/2026
02225574-0016